THE MEANING OF EVOLUTION

EVOLUTION

●

Charles Darwin, 1809–1882, wedding portrait done in 1841.

For Barbara

CONTENTS

ILLUSTRATIONS

PREFACE

This book began as a twelve-page essay on the history of the word "evolution," which had been solicited for a volume on key terms in evolutionary biology. I had assumed my efforts would be perfunctory, since historians of science had previously established two distinct and nonoverlapping meanings for "evolution." In the seventeenth century the word had been recruited to refer to the embryological theory of preformationism, according to which the embryo from the very beginning existed as a miniature adult that simply unfolded or "evolved" during gestation. The second meaning, assumed to have been given currency by Herbert Spencer in the 1850s, referred to species descent with modification. Historians have understood these two usages of "evolution" to be quite separate in meaning, like the "bark" of the dog and the "bark" of the tree; and to link them would have been comparable to supposing that a tree might bite you. However, as I slipped into the research, I became ever more wary of trees. I now believe these terms are historically joined through a process not unlike evolution itself: the older embryological idea gradually became transformed over two centuries into the more familiar one referring to species change, while yet retaining vestiges of its past.

The historical missing link between the two meanings of evolution was a theory that had been thought, according to the usual historiography, to have achieved real significance only at the end of the nineteenth century, when it had been crudely used by that archprogressionist and bête noire of all good Darwinians, Ernst Haeckel. Haeckel had made the hinge of his evolutionary theory the biogenetic law, the dogmatically driven doctrine (or so it was portrayed) that the embryo retraced the same morphological steps the species went through in its evolutionary development, that ontogeny recapitulated phylogeny. The established orthodoxy had recapitulation theory as that sign of Cain by which ideologically conceived science, like Haeckel's, could be distinguished from the legitimate ancestor of modern neo-Darwinism.

Something there is about a historical orthodoxy that wants it down. I had been laboring over a book on Haeckel and was led to understand his form of German Darwinism, despite all of its deliciously romantic features, not as an impostor but as strongly related to Darwin's own child. That British offspring of genius, I began to suspect, might also have carried the genes, as it were, for recapitulation. Now returning to Darwin's historical development with this as a possibility—and motivated as most scholars are to kill the king and marry the queen—I believed I had discovered that Darwin's theory, from its conception to its maturation, pulsed to the rhythms of that ever-fascinating principle.

The first half of this book, then, attempts to show how the original meaning of evolution became transformed, passing through the principle of recapitulation, finally to refer to species descent. I use the term "evolution" itself, and its various cognate and derived forms in the several European languages, as the marker by which to aid in this historical reconstruction. The second half of the book argues that Darwin had been not merely an advocate of recapitulation but that the principle was such an intrinsic feature of his conception one could not take the real measure of his theory without understanding that he, like several of his important predecessors in the tradition of transcendental morphology, had conceived embryological evolution and species evolution as really two aspects of the same process: both kinds of evolution involved a gradual morphological change through a sequence of the same type patterns. The evidence for Darwin's reliance on an embryological model of evolution now seems to me so strong and the use of recapitulation so flagrant, the real puzzle becomes why so many historians have argued vehemently just the opposite case. I believe the answer lies along the darker byway of ideology, and in the conclusion of this book I attempt to feel my way down that treacherous path.

My research for this project was funded by the National Science Foundation and the National Endowment for the Humanities. The staff of the Manuscript Room of Cambridge University Library offered considerable help with manuscripts and gave permission to quote from unpublished material. I am indebted as well to the Special Collections Department of the University of Chicago Library for furnishing access to the rare volumes necessary for this study. Elisabeth Lloyd and Evelyn Fox Keller commissioned the essay that stimulated my investigations, and they offered guidance on that initial pass. Susan Abrams, executive editor of the University of Chicago Press, and David Hull, series editor, encouraged me to publish my research as a small book. Charles Dinsmore, Lynn Nyhart, Brian Ogilvie, Michael Ruse, Marc Swetlitz, Phillip Sloan, and Tracy Teslow quite generously read the manuscript and provided much-needed advice. Peter Bowler kindly attempted to warn me away from what he perceived as historiographic disaster. And students in my seminar (fall 1990) on Darwin's *Origin of Species* were always ready, in the tender Chicago manner, to incise the boil of a bad idea, while salving the sore spot with many creative alternatives. I am grateful to these readers for their efforts.

1
THE NATURAL HISTORY
OF IDEAS

Who today would claim to be a historian of ideas? The field upon which Lovejoy played seems now too fanciful, too removed from the actual concerns of the individuals whose ideas he joined with Platonic hoops of reason. Even to admit to intellectual history, where passing attention is paid particular circumstances, would be risky in a department where "real" history is done. A colleague of mine, a Tudor-Stuart historian, dismissed intellectual history with the remark that the people he studied didn't think. Historians of science, when they reside in history departments, may be indulged. They do, after all, write about thinkers. But the staid and minute practices of historians of science have become more troubled, and their older ways begin to appear indefensible. From the American side, the forces of the social historians press against work that fails to consider institutional context and economic constraints; on the other side of the Atlantic, they cultivate disdain for historical analyses that ignore class jealousies and the kind of Molièrean interests that seem to explain everything. These several factors used to be classified as part of external history; they might buffet about the internal organs of science but could do no real damage. Certainly for historians dealing with their subjects now, it is hard to tell the inside from the outside—especially when one considers (not that the social constructionists would) that a scientist might have a strong interest in achieving a true theory; for a true theory might bring fame, power, and the love of interesting people. No one today would deny that social factors often form the environment for the development of scientific ideas. The mistake is to assume a priori that they must always be the most important causes of thought. But whether or not social factors turn out to be significant in particular cases, the historian must first come to understand, with some intimacy, the scientific theories, ideas, and observations for which an explanation is sought. And in this pursuit, he or she might well consider the model provided by natural historians, whose practices suggest how intellectual history and history of ideas might be revivified.

Before Darwin, natural historians had systematized relations among animals; they traced out those homologies that united apparently disparate groups into common types. And as the tale I will tell indicates, these early natural historians even thought of the relations they discussed in genealogical terms—as if one animal group had given birth to another. Darwin, though, interpreted these connections as dynamic, as real genealogical relations of descent. He then engaged his various devices, chiefly natural selection, to give them further causal account. The historian of science might look to Darwin and his predecessors as a guide. For the historian can as well trace the genealogy of ideas and theories, using similar intellectual structures as suggestive of possible relations of descent. But even in preliminary studies, when the focus is broad, the historian must do more. He or she must be careful not to mistake analogies for homologies, not to assume that because one set of ideas is similar to another it must have descended from that other. The historian must demonstrate that a particular scientist met with, corresponded with, or read the books of the person whose ideas seem to be the progenitors; this must be done in order to establish the firmer ground of probability for real genealogical relations. Further, the historian must, insofar as possible, isolate those intellectual exigencies, as well as those social and psychological pressures, that render comprehensible the selections made by a scientist from the range of possible ideas. Without the possibilities narrowed and the probabilities fixed, history will be haunted by the zeitgeist. Older historians of ideas often failed to exorcise this spirit and mistook intellectual analogies for homologies. Recent, socially constructivist historians have taught us not to be afraid, even when we should.

In what follows, I wish to undertake an examination that appears to be in the hoary tradition of the history of ideas. My effort will be to trace out the changing significance of the term "evolution." But I will use the word as an index to probe the vitality of a larger set of ideas and as a recurrent test for the real interactions and intellectual connections among naturalists from the seventeenth century through Darwin's lifetime. The genealogy of "evolution" will reveal, I believe, the bloodlines of Darwin's own emerging theory, showing that theory, oddly, to be a product of nineteenth-century thought, even conceived through dark relations with transcendental morphology. In attempting to lay bare these deeper connections, I will distinguish, in the next two chapters, several moments in the gradual alteration of the meaning of "evolution," namely: its initial use to describe the embryological condition; its transformation in the transcendental principle of recapitulation—the idea that the embryo of a higher organism passes through the adult forms of lower organisms; and its new life as a term for species change. In chapter 4 I will describe the early British discussions of embryological recapitulation and species evolution, and sketch Darwin's own initial considerations of these topics. In chapter 5 I will try to show how the earlier set of ideas that "evolution" indexes became the structure upon which Darwin formed his own theories of species alteration. More specifically, I will indicate how, as I believe, embryological development became for Darwin a model of descent, infusing his conception precisely with those attributes, especially notions of progress, usually thought to characterize only non-Darwinian, romantic theories of evolution in the nineteenth century. The Darwin that emerges from this study will appear decidedly more venerable than that rejuvenated evolutionist who has been injected by some historians with the monkey glands of a modern scientific ideology, a treatment I will discuss in the concluding chapter.[1]

1. Two previous essays of considerable interest have traced the meaning of "evolution," one by Thomas Henry Huxley ("Evolution," *Encyclopaedia Britannica,* 9th ed., vol. 8, 1878) and the other by Peter Bowler ("The Changing Meaning of 'Evolution,'" *Journal of the History of Ideas* 36 (1975): 95–114). Both Huxley and Bowler, like many others, note what seem to be the two distinct biological uses for the term: initially, to refer to the particular embryological theory of preformationism; and later, to characterize the general belief that species have descended from one another over time. Bowler maintains that in the 1850's a link was forged between a species-transforming meaning of "evolution" and an embryological-developmental meaning of the term, the latter having been suggested by von Baer. However, he argues that in Darwin's case, "it is not easy to connect the use of the term with the idea of progression and hence with the embryological analogy. The transmutation of species need not be a progressive process, and Darwin's theory was certainly not developed as an explanation of progression" (p. 101). The argument of my essay runs exactly against the grain of Bowler's conclusions and the comparable ones of Ernst Mayr, Dov Ospovat, and Stephen Jay Gould, as I will explain below.

2

EVOLUTION VS. EPIGENESIS IN EMBRYOGENESIS

The Latin infinitive *evolvere* means "to unfold or disclose." *Evolutio,* its substantive form, refers to the unfolding and reading of a scroll, as in Cicero's *De finibus:* "Quid poetarum evolutio voluptatis affert?" (What pleasure does the reading of the poets provide?).[1] The *Oxford English Dictionary* suggests that "evolution" was first used in a biological context by an anonymous English reviewer in 1670; subsequently, and more importantly, Albrecht von Haller employed the Latin form, in 1744, for the same purpose as the Englishman, that is, to characterize the preformationist embryology of the Dutch entomologist Jan Swammerdam (1637–80).

In his posthumously published *Historia insectorum generalis* (1685), Swammerdam had argued that an insect grub was generated from the semen of the female (stimulated by the semen of the male) and that the adult form lay already encased in the embryonic larva (see Fig. 1), requiring only that its outer skin be shed and its preexisting internal parts be augmented and expanded.[2] He measured this theory of preformation against William Harvey's (1578–1657) notions of metamorphosis and epigenesis. In his *Exercitationes de generatione animalium* (1651), Harvey isolated two distinct modes of ontogenetic (that is, individual) development: the process *per metamorphosin,* in which all the organs of an in-

---

1. Cicero, *De Finibus,* 1.7.25.
2. Jan Swammerdam, *Historia insectorum generalis,* translated from the Dutch by H. Henninius (Holland: Luchtmans, 1685), pp. 44–45.

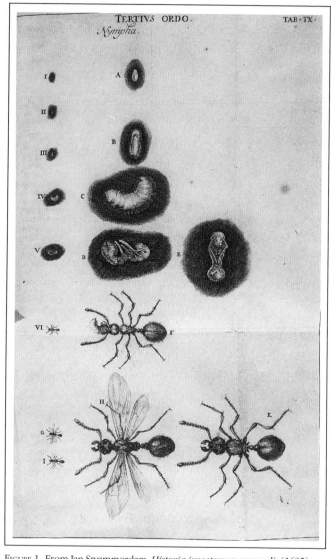

FIGURE 1. From Jan Swammerdam, *Historia insectorum generalis* (1685). Figures VD and VE show an adult ant encapsulated in the larva.

sect became transfigured simultaneously, when, for instance, the caterpillar was transformed into a butterfly; and the process *per epigenesin,* in which the organs of the higher animals developed sequentially, changing gradually from an amorphous, homogeneous condition to an articulated, heterogeneous state.[3] Swammerdam disdained these Aristotelian processes, requiring as they did a nonexperiential "plastic force [*vis plastica*]" to guide them. His many careful microscopical and dissectional studies of insects and frogs convinced him that for every animal "certainly the forms and articulations of all the parts preexist in the fecund semen."[4] Such preexistence, Swammerdam suggested, did not mean that the semen carried an actual physical miniature—though his later readers so interpreted him. He held, rather, that the generative material was predelineated "in ideas and types according to a rational similitude." This doctrine implied that "the entire human race already existed in the loins of our first parents, Adam and Eve, and that, for this reason, all of human kind has been damned by their sin."[5] Theological larvae have always danced over the several theories of evolution.

The English reviewer, in the *Philosophical Transactions of the Royal Society,* observed that when Swammerdam referred

3. William Harvey, *Exercitationes de generatione animalium* (London: DuGaidianis, 1651), p. 121.

4. Swammerdam, *Historia insectorum,* p. 45: "typos certe & . . . omnium partium delineationes in semine foecundo praexistere."

5. Ibid., p. 46. The theory of "preformation," for which Swammerdam argued, has been distinguished by some historians from the notion of "preexistence," which suggests that each preformed embryo carries a seed containing another embryo, which in turn carries a seed containing yet another embryo, so that all descendants of the original progenitor preexist in a cascading series of encasements. See, for instance, Jacques Roger, *Les Sciences de la vie dans la pensée francaise du XVIIIe siècle,* 2d ed. (Paris: Armand Colin, 1971), pp. 325–26. It is certainly true that the theory of embryonic encasements received its more elaborate development in the eighteenth century, especially by Bonnet (see below). The above passage from Swammerdam, however, indicates that the simple logic of preformationism was appreciated from the beginning. I will therefore use the terms "preformation" and "preexistence" with some inattention to the distinction.

to the "change" that insects underwent, "nothing else [is] to be understood but a gradual and natural Evolution and Growth of the parts."⁶ The great Swiss anatomist Albrecht von Haller (1708–77) similarly characterized Swammerdam's conception. Haller weighed rival theories of embryological change in his commentary on the *Praelectiones academicae* (1744) of his own famous teacher at Leiden, Hermann Boerhaave (1668–1738). Aristotle, Harvey, and a few Italians proposed epigenesis, which taught that "the parts of animals are successively generated out of fluid according to certain laws." However, "the theory of evolutions [*evolutionum theoria*] proposed by Swammerdam and Malpighi obtains almost everywhere."⁷ The theory of evolutions holds that "all the viscera, muscles, and remaining solid parts have already existed in the first beginnings of the invisible human embryo, and that they have at length successively become apparent in those places where they have been slowly dilated by an influxing humor and have become a visible mass."⁸ Perhaps the Latin alternate title of Swammerdam's book, *Biblia naturae,* inspired the poet in Haller (and the English reviewer) to coin this unscrolling usage for *evolutio.* Haller observed that by the mideighteenth century, the theory of evolutions had generated two camps, the ovists, like Swammerdam, who "teach that some sort of germ or perfect human machine exists in the egg," and the spermists, like Boerhaave, who believed that "man preexists in the little worm and that . . . the fabric of the whole body has been delineated in the earliest embryonic stage and that it is expanded by heat and reabsorbed humor."⁹

6. Anon., "Review of *Historia Generalis Insectorum,*" *Philosophical Transactions of the Royal Society* 5 (1670): 2078–80; quotation from p. 2078.

7. Albrecht von Haller, in Hermann Boerhaave, *Praelectiones academicae,* edited with notes by Albertus Haller, vol. 5, part 2 (Göttingen: Vandenhoeck, 1744), p. 489: "evolutionum theoria fere ubique obtinet a Swammerdamio . . . & Malpighio proposita."

8. Ibid.

9. Ibid., p. 490.

THE MEANING OF EVOLUTION

The Morphological Construction and
Ideological Reconstruction
of Darwin's Theory

•

Robert J. Richards

Science and Its Conceptual Foundations
David L. Hull, Editor

The University of Chicago Press
Chicago and London

ROBERT J. RICHARDS is professor in the departments of History, Philosophy,
and Psychology at the University of Chicago.

The University of Chicago Press, Chicago 60637
The University of Chicago Press, Ltd., London
© 1992 by The University of Chicago
All rights reserved. Published 1992
Printed in the United States of America

01 00 99 98 97 96 95 94 93 92   5 4 3 2 1

ISBN (cloth): 0-226-71202-8
ISBN (paper): 0-226-71203-6

Library of Congress Cataloging-in-Publication Data

Richards, Robert J., date.
    The meaning of evolution : the morphological construction and
ideological reconstruction of Darwin's theory / Robert J. Richards.
        p.   cm. — (Science and its conceptual foundations)
    Includes bibliographical references and index.
    I. Evolution (Biology)   I. Title   II. Series.
QH366.2.R52   1992
575—dc20   91-19017
                                                   CIP

♾ The paper used in this publication meets the minimum requirements
of the American National Standard for Information Sciences—Permanence
of Paper for Printed Library Materials, ANSI Z39.48-1984.

ALBERTUS DE HALLER

Huic lex summa fuit Naturæ voce doceri
Huic dominæ doctas subdidit artis opes ;
Ingenuus veri vel ab hoste nitentis amicus
Censor & erroris candidus ipse sui.

P. G. Werlhof.

F. J. Handmann Pinx.

P. F. Tardieu Sculp.

Chez MARC-MICHEL BOUSQUET et Comp.^e 1757.

FIGURE 2. Albrecht von Haller, 1708–1777. "The supreme rule for him was to be instructed by the voice of nature, and he submitted to this mistress the learned power of his skill; he was a noble friend of truth as well as a friend in a struggle with a foe, and he was a most severe and forthright critic of his own error." Wellcome Institute Library, London.

Haller, in his commentary on Boerhaave, momentarily cast his lot with the Aristotelians. Earlier he had held preformationist views. Now he thought the conceptual balance dipped toward epigenesis, especially as one considered the apparent generation of animal and plant parts out of fluid antecedents, the regeneration of limbs in lower creatures, and the gradual development of the fetal heart. Yet for him the theoretical situation remained unsettled. And as his respect grew for the Newtonian mechanical philosophy, which brooked no theologically suspect self-formative powers, he again bethought his position. In a brighter and theologically safer Newtonian light, he undertook an experimental reexamination of embryological theory. His extended researches on fertilized chicken eggs, in which he traced more articulated embryonic structures to their supposed folded and translucently invisible earlier stages, provided him the compelling evidence he needed for a return to his initial conviction, a theory of embryological evolution.[10] During his research and in the solidification of his theoretical ideas, Haller received constant encouragement from his countryman and friend Charles Bonnet (1720–93), whose discovery of parthenogenesis in aphids and microscopical observations of the aphid imago just beneath the skin of the grub, coupled with his degenerating vision and accelerating religious enthusiasm, undoubtedly disposed him also to champion preformationism.

Bonnet is chiefly responsible, in the eighteenth century, for transposing individual into species evolution. In his highly influential *Considerations sur les corps organisés* (1762), he advanced "evolution [*l'Evolution*] as the principle that better conforms to the facts and to sound philosophy."[11] Later, in his more speculative *La Palingénésie philosophique* (1769), he elaborated his complementary, though tentative, doctrine of

10. For a thorough discussion of Haller's theories, see Shirley Roe, *Matter, Life, and Generation: 18th-Century Embryology and the Haller-Wolff Debate* (Cambridge: Cambridge University Press, 1981).

11. Charles Bonnet, *Considerations sur les corps organisés,* 2 vols. (Amsterdam: Marc-Michel Rey, 1762), 1: vi.

FIGURE 3. Charles Bonnet, 1720–1793, and the great chain of being; engraving done in 1802 for Thornton's *Sexual System of Linnaeus*. Wellcome Institute Library, London.

emboîtement, or encapsulation, to produce a general theory
of "the natural evolution of organized beings [d'Evolution
naturelle des Etres Organisés]."[12] Bonnet, fervent if heterodox Calvinist that he was, believed that God had originally
created a plenitude of germs, each encapsulating a miniature
organism that in turn carried germs containing yet more
homunculi and their germs, enough to reach the Second
Coming. Every miniature bore all the adult species traits,
though it required the maternal (or other ontogenetic) environment to individuate its form.[13] With each of the great
catastrophes the world had undergone, of which Genesis recorded the aftermath of the most recent, animals and plants
became extinct, though not their germs. Those saved rem-
nants were preserved to flower anew. Each of the encapsulated lines, Bonnet supposed, might contain animals or plants
of more than one species. After each catastrophe, then, another set of more perfectly developed species could evolve
from the germs of the old,[14] such that there might be "a continued progress of all species, more or less slowly, towards a
higher perfection."[15] Bonnet thus extended the idea of embryological evolution to construct a theory of the unfolding of
preformed and ever more perfect species. Though his "natural evolution" appears only remotely related to Darwin's conception of contingently branching species, even Thomas

12. Charles Bonnet, *La Palingénésie philosophique, ou Idées sur l'état
passé et sur l'état futur des êtres vivans,* 2 vols. (Geneva: Philibert et Chiroi,
1769), 1: 250.

13. Bonnet, *Considerations sur les corps organisés,* 2: 462–63.

14. Ibid., pp. 253–55. Bonnet believed there would be at least three great
revolutions on the globe: the first world coming directly from the hand of the
Creator; the world renewed after the Mosaic flood; and the world coming to
be with the resurrection of the body (p. 253). He allowed, however, that if
one wished to propose many such revolutions, with each leading to greater
perfection of species, he would have no objection (p. 254).

15. Ibid., p. 204: "un progrès continuel & plus ou moins lent de toutes les
Espèces vers une Perfection supérieure."

Henry Huxley detected in it "no small resemblance to what is understood by 'evolution' at the present day."[16]

Shortly after Haller published his new theory of embryological evolution, it was attacked by the young German physician Caspar Friedrich Wolff (1734–94). In his doctoral dissertation *Theoria generationis* (1759), Wolff defended an epigenetic theory against Haller's "mechanistic medicine," which explained "the body's vital functions from the shape and composition of its parts."[17] Braced by the empirico-rationalistic traditions established at Halle in the previous generation by Christian Wolff,[18] he undertook a study of the vascular system of the embryonic chick, which convinced him that the animal's vessels formed out of homogeneous matter under the aegis of "a principle of generation, or essential force [*vis essentialis*], by whose agency all things are effected."[19] With his *Theorie von der Generation* (1764), Wolff also brought Bonnet within his sights, and feeling relieved from the constraint produced by his respect for Haller,[20] he rather delighted in exploding the confusions Bonnet had set loose concerning Haller's conception of pre-

16. Thomas Henry Huxley, "Evolution," *Encyclopaedia Britannica,* 9th ed. (1878), 8: 745. Huxley's description of Bonnet's ideas stimulated C. O. Whitman to undertake an extensive historical analysis, which is still of considerable interest. See Charles Otis Whitman, "Bonnet's Theory of Evolution," and "The Palingenesia and the Germ Doctrine of Bonnet," in *Biological Lectures Delivered at the Marine Biological Laboratory of Wood's Hole,* ed. C. O. Whitman (Boston: Ginn & Company, 1895), pp. 225–40, 241–72.

17. Caspar Friedrich Wolff, *Theoria Generationis* (Halle: Hendelianis, 1759), p. 124.

18. Christian Wolff was unrelated to Caspar Friedrich Wolff. For a discussion of Christian Wolff's ideas, see Robert J. Richards, "Christian Wolff's Prolegomena to Empirical and Rational Psychology: Translation and Commentary," *Proceedings of the American Philosophical Society* 124 (1980): 227–39.

19. Caspar Friedrich Wolff, *Theoria Generationis,* p. 106.

20. See Roe's account of the cordial debate between Wolff and Haller in her *Matter, Life, and Generation.*

delineation.[21] His general, theoretical criticism of both authors, however, was quite direct and to the mark: "one finds nothing in nature which would be similar to an evolution [Man findet nichts in der Natur welches einer Evolution ähnlich wäre]," so that the probability of the hypothesis was greatly reduced.[22] Moreover, the hypothesis assumed that an evolved phenomenon (*entwickeltes Phänomen*) was really a prefabricated miracle; and even though such a miracle might have originally been performed at the creation, the notion simply embarrassed the quite contrary and established "concept we have of . . . a living nature which undergoes countless changes through its own power."[23]

By the end of the eighteenth century, the embryological work of Wolff added Germanic experimental thoroughness and theoretical comprehensiveness to the rising tide of epigenetical studies washing every shore: for example, in England, John Needham (1713–81) made microscopical observations of the spontaneous generation of infusoria—certainly evidence against preformationism—while his sometime collaborator in France, Georges Leclerc, Comte de Buffon (1707–88), advanced other arguments against the Hallerian position, supplying in its stead a complex epigenetic theory of embryological development. The older evolutionary theory thus succumbed to better microscopes, to the decline of interventionist theology, and to the new enthusiasm for independent disciplines and special formative principles.

Preformationism, however, did not go extinct immediately or completely. Georges Cuvier (1769–1832), the Baron of French science at the Museum d'histoire naturelle and the very model of the nineteenth-century scientist, held on to the old evolutionary theory because German plastic principles were both theologically distasteful and contrary to the idea of

21. Caspar Friedrich Wolff, *Theorie von der Generation* (Berlin: Birnstiel, 1764), pp. 97–135.
22. Ibid., pp. 40–41.
23. Ibid., p. 73.

unitary *embranchements* in the animal kingdom.[24] For Cuvier, though, the theory was but a lingering thought on which he never insisted. The conception, however, did survive by itself undergoing a kind of evolution. At the turn of the century many embryologists detected in the development of the fetus, not the expansion of the already-formed adult of that species, but the serial unfolding of the adult forms of more primitive species. The embryo seemed to recapitulate sequentially the hierarchy of species below it. As Etienne R. A. Serres (1786–1868) expressed it in 1824:

> Embryos, therefore, are not, as it has been imagined, the miniature of the adult animal. Before they arrive at their permanent forms, their organs traverse a multitude of fugitive forms, beginning with the most simple, if we look to their starting point. Remarkably, those embryonic forms in the superior classes frequently repeat the permanent forms of the inferior classes.[25]

Through the turn of the century, anatomists thus began to slide from the notion of the embryo as a miniature adult of its own species to that of the embryo as a sequence of miniature adults of lower species. And the term "evolution" tracked this conceptual movement. It came to refer to this sort of progressive embryological development and then, as the theory of recapitulation matured, to progressive species development. Indeed by the 1830s, the word "evolution" had shifted 180 degrees from its original employment and was used to refer indifferently to both embryological and species progression. This change can be seen in the passing observation, made in 1833, of Etienne Geoffroy Saint-Hilaire (1772–1844), a colleague and supporter of Lamarck and mentor of Serres: "Of the two theories of the development of organs, one supposes

24. Georges Cuvier, *Histoire des progrès des sciences naturelles depuis 1789 jusqu'a ce jour,* 4 vols. (Paris: Baudouin Frères, 1829), 1: 240–41.

25. Etienne Reynaud Augustin Serres, *Anatomie comparée du cerveau,* 2 vols. (Paris: Gabon, 1824–27), 1: xv–xvi.

the preexistence of germs and their indefinite *emboîtement,* the other acknowledges their successive formation and their evolution [*leur évolution*] in the course of ages."[26]

In the next three chapters, I will follow the development of evolutionary theory from its early formation in English, French and German hands to its use by Darwin in the construction of his theory. A constant guide in this tracking will be an author's use of the term "evolution" and its semantic equivalents. At the turn of the eighteenth century, "evolution" (in its Latin, English, French, and German cognate forms) acquired new semantic properties and linguistic equivalents arose. So in German, *Entwickelung* often performed the same functions as the Latin *evolutio;* and in English the "theory of development" was used interchangeably with the "theory of evolution." What remained constant, however, was that the same word, whether it was *evolutio,* "evolution," *Entwickelung,* or "development," would be used by an author to refer both to embryological transformation and to species transformation.

26. Etienne Geoffroy Saint-Hilaire, "Le degré d'influence du monde ambiant pour modifier les formes animales," *Mémoires de l'Académie royale des sciences de l'Institut de France,* 2d ser. 12 (1833): 63–92; quotation from p. 89.

3

THE THEORY OF EVOLUTIONARY RECAPITULATION IN THE CONTEXT OF TRANSCENDENTAL MORPHOLOGY

The renowned embryologist and later opponent of Darwinian theory Karl Ernst von Baer (1792–1876) indexed the shifting meaning of "evolution" in his critical rejection of recapitulation theory. In his celebrated *Entwickelungsgeschichte der Thiere* (1828), he referred to his earlier (1823) academic disputation, wherein he argued that "the law proclaimed by naturalists was foreign to nature, namely that 'the evolution which each animal undergoes in its earliest period corresponds to the evolution which they believe to be observed in the animal series [evolutionem, quam prima aetate quodque subit animal, evolutioni, quam in animalium serie observandum putant, respondere].'"[1] By the 1820s, the principle of recapitulation, against which von Baer's disputation was directed, had gained a strong hold on the imagination of German biologists; and it guided their investigations as they moved through realms of transcendental morphology.[2]

1. Karl Ernst von Baer, *Entwickelungsgeschichte der Thiere: Beobachtung und Reflexion* (Königsberg: Bornträger, 1828), pp. 202–3.
2. The term "biology" was first used by Karl Friedrich Burdach (1776–1847), teacher of von Baer at Tartu and later his colleague at Königsberg, to refer to the study of man from a zoological and physiological perspective. The term received its more modern meaning at the hands of Gottfried Reinhold Treviranus (see below). In his multivolume treatise *Biologie, oder*

Early Recapitulation Theorists

Despite the predominantly German provenance, an Englishman, in the late eighteenth century, first shaped the idea of recapitulation into identifiable form. The great autodidactical physiologist John Hunter (1728–93) has the best claim on priority, a claim which made deployment of the principle in England during the next century a rather complex affair, especially in the case of Richard Owen. In descriptions accompanying Hunter's sketch of a chick embryo—one of the manuscripts that escaped the flames set by his plagiarizing son-in-law—he observed that "if we were to take a series of animals from the imperfect to the perfect, we should probably find an imperfect animal corresponding with some stage of the most perfect."[3] Somewhat later, in 1793, Karl Friedrich Kielmeyer (1765–1844)—fellow student and tutor of Cuvier

---

Philosophie der lebenden Natur, he introduced the new term and thereby helped establish a field of inquiry: "The objects of our research will be the different forms and manifestations of life, the conditions and laws under which these phenomena occur, and the causes through which they have been effected. The science that concerns itself with these objects we will indicate with the name biology [*Biologie*] or the doctrine of life [*Lebenslehre*]." In the same year, 1802, Lamarck made a similar application of the term. See Gottfried Reinhold Treviranus, *Biologie, oder Philosophie der lebenden Natur,* 6 vols. (Göttingen: Johann Friedrich Röwer, 1802–22), 1: 4. See also Richard Burkhardt, "Biology," *Dictionary of the History of Science,* ed. W. F. Bynum et al. (London: Macmillan, 1981), p. 43. I will discuss the meaning of "transcendental morphology" in the text below.

3. This passage was quoted by Owen in his redaction of the manuscripts of Hunter. See [Richard Owen], "Introduction," *Descriptive and Illustrated Catalogue of the Physiological Series of Comparative Anatomy Contained in the Museum of the Royal College of Surgeons in London,* vol. 5: *Products of Generation* (London: Taylor, 1840), p. xiv. Owen cited a similar passage from Hunter's "Croonian Lecture" of 1782. Joseph Henry Green, Owen's mentor, remarked in his Hunterian lecture of 1840 that Wolff and Meckel had established the law "already anticipated by Hunter, that the progressive phases of the embryo correspond to the abiding forms, which are preserved in the total organism of animated nature, as typical of its gradative evolution." See Joseph Henry Green, *Vital Dynamics: The Hunterian Oration before the Royal College of Surgeons in London, 14th February 1840* (London: Pickering, 1840), p. 39.

at the Karlsschule near Stuttgart—alluded vaguely to comparable general stages in the early embryogenesis of men and birds and remarked that sense organs appeared in the individual "almost in the same order" as in the series of lower organisms.[4] Though the principle remained undeveloped, at least in the small number of his publications, Kielmeyer did insist on a fundamental constituent of recapitulation, which would govern its use by virtually every subsequent zoologist (including Darwin), namely, the assertion that the laws governing the evolution of species were the same as those governing the evolution of embryos. As he put it, in his Kantian effort to unify the understanding of nature: "the force by which the series of species has been brought forth is one and the same in its nature and laws as that by which the different developmental stages [in embryogenesis] are produced."[5] Kielmeyer moved to a professorate in chemistry at Tübingen in 1796, where he had as a colleague Johann Heinrich Autenrieth (1772–1835).

The year Autenrieth arrived at Tübingen, 1797, he published his *Supplementa ad historiam embryonis humani,* an exacting empirical study of the growth of the human fetus. In the course of his sequential observations, he also noted the comparisons to be made between "the diverse stages in the evolution [*evolventis status*] of the [human] embryo and

4. Karl Friedrich Kielmeyer, "Ueber der Verhältnisse der organischen Kräfte unter einander in der Reihe der verschiedenen Organisation" (1793), reprinted in *Sudhoffs Archiv für Geschichte der Medizin und der Naturwissenschaften* 23 (1930): 247–67; quotation from p. 261.

5. Ibid., p. 262. Kielmeyer published virtually nothing during his lifetime, so his influence could only be channeled through his students and colleagues (like Cuvier). From the passage just cited, it would appear he had a prototransformationist conception, something also suggested by a line quoted in the dissertation of his student Johann Tritschler: "the series of animals is continuous and directly ascending, with each member distinguished by its grade of evolution and heterogeneity of its organs [seriem animalium esse continuam, recta ascendentem quarum membra modo different gradu evolutionis et heterogeneitatis organorum]." The passage from Tritschler is given in Felix Buttersack, "Karl Friedrich Kielmeyer," *Sudhoffs Archiv für Geschichte der Medizin und der Naturwissenschaften* 23 (1930): 236–46.

the constant forms of the inferior animals." He even suggested that certain internal features of lower creatures would
"be seen less changed in the African adult than in the European adult."[6] Autenrieth used *evolutio* throughout his tract in
full knowledge of its history but with the intention to synthe-
size in that term Haller's notion of mechanical determination
with Blumenbach's conception of a more vitalistic *nisus formativus.*[7] The chief German proponents of the recapitulation
principle looked back to Kielmeyer and Autenrieth as the
original authorities. Though von Baer cautiously avoided
mentioning his opponents by name, he must certainly have
had these influential naturalists in mind when mounting his
opposition to the idea of parallel evolutions. Internal evidence suggests, however, that his primary targets must have
been contemporary rivals who more extensively used recapitulation theory, namely Oken, Tiedemann, Treviranus, and
Meckel.[8]

6. Johann Heinrich Autenrieth, *Supplementa ad historiam embryonis humani* (Tübingen: Heerbrandt, 1797), p. 24: "Plura ad comparandos diversos
embryonis se evolventis status cum constantibus inferiorum animalium
formis in interna structura partim evidentissima, & quaedam, quae in adulto
Afro minus, quam adulto Europaeo ex reliquiis embryonis mutata videntur,
progressus opusculi exhibebit."

7. Ibid., pp. 2–2. Blumenbach took a strong though, for the period, an obvious stand against Hallerian evolutionary theory and defended epigenesis as
the reasoned alternative. An epigenetical approach to embryonic development required the postulation of a productive force, a *nisus formativus* or
Bildungstrieb, which he defined in rather Molièran fashion: "It is a drive
[*Trieb*] completely distinct from all merely mechanical formative powers
(such as that which produces crystallizations in the organic realm). It is able
to modify the many sorts of organizable seminal material in many different
yet goal directed ways and combine them into determinate forms." Blumenbach recognized that strictly speaking this causal power remained a *qualitas
occulta;* yet, as he maintained, we could order and explain other organic
phenomena, such as growth and repair, through its postulation. See Johann
Friedrich Blumenbach, *Handbuch der Naturgeschichte,* 12th ed. (Göttingen:
Dieterich'schen Buchhandlung, 1830), §9, p. 15–17.

8. Other German zoologists, of course, also advanced recapitulation theory in the early nineteenth century. Carl Gustav Carus, for instance, followed
Meckel and Oken in the promotion of the "most important of [the facts of

Naturphilosophie and Transcendental Morphology

The various biological ideas of the aforementioned German naturalists are usually taken as expressions of "transcendental morphology," also called "ideal morphology" and "transcendental anatomy." The seeds dispensed in the proliferating growth of transcendental morphology reached British shores in the 1820s and came to full, if appropriately reserved, flower in the work of Richard Owen; Darwin nurtured them in different soil, transforming them and fitting them to serve in his evolutionary theory of species change. Transcendental morphology itself grew in the wild luxury of German philosophy of nature, *Naturphilosophie.* In order to understand better the meaning of transcendental morphology then, and ultimately theories of recapitulation and their relation to species evolution, we should spend a few moments inspecting the common context that inspired the romantic excesses of biologists like Oken, the literary exuberance of writers like Coleridge, and the steadier reflections of experimentalists like von Baer.

Naturphilosophie derives from several sources, but its intellectual core grew from Kant's critical philosophy, Schelling's transcendental idealism, and Goethe's developmental morphology.[9] In his First Critique, the *Kritik der reinen Ver-*

---

anatomy], namely that the development [Entwicklung] of individual animals represents the progressive stages of the animal kingdom." Carus cautioned, however, that the organs of the fetus of higher animals were not identical to those of the lower, but only of the same general form. Moreover, the animal kingdom displayed different developmental series, so the fetus of a given higher class might not traverse the same series as the fetus of a different class. See Carl Gustav Carus, *Lehrbuch der Zootomie* (Leipzig: Gerhard Fleischer the Younger, 1818), pp. 667–70; also see n. 77, below. Carus's *Lehrbuch* was introduced into English circles by Joseph Henry Green, the mentor of Richard Owen (see chap. 4, below).

9. For a useful collection of short articles that touch on various aspects of *Naturphilosophie,* see Andrew Cunningham and Nicholas Jardine, eds., *Romanticism and the Sciences* (Cambridge: Cambridge University Press, 1990). Lynn Nyhart exquisitely reconstructs the fate of *Naturphilosophie* and German morphology during the second half of the nineteenth century. See

nunft (1781), Immanuel Kant (1724–1804) argued that the immediate, manifold deliverances of sensation, which presumably stemmed from an external, noumenal world beyond experience, could not provide the structure that our awareness of the everyday, phenomenal world actually exhibited. In order to account for the organization of the phenomenal world, we had to assume that our understanding imposed certain modes of thought upon sensation so as to constitute the necessary structural features of our experience of that world. The coherence of our perceptions, the unity of objects perceived, their causal interrelatedness, and even the spatial and temporal modes of experience could only be explained, Kant concluded, if we supposed that the mind itself provided organizing categories and forms. Kant's Copernican revolution, whereby he made, as it were, the observer supply the motions imputed to the external world, could also account for the validity of the fundamental principles of mathematics and natural science. We can always be sure that the next triangle we observe will display Euclidian properties because we ourselves implicitly structure our perceptions according to the rules of that geometry. Objects causally impinging on others will necessarily receive a reciprocal force, since that is the way we must think of them. Kantian transcendental philosophy thus provided justification for the universality and necessity exhibited by the laws of Newtonian mechanics. The epistemological situation of biology, however, was somewhat different.

In the *Kritik der Urteilskraft* (1790), the Third Critique, Kant maintained that while we necessarily imposed mechanistic conceptions of causality in the constitution of experience, these proved insufficient in our understanding of the phenomena of life. The harmonious skein of contingent principles, whereby we attempted to understand the organization and development of creatures, required us, because of the

---

her *Before Biology: Animal Morphology and the German Universities, 1850–1900,* forthcoming.

exigencies of human judgment, to assume that living beings expressed a purposeful design, an *intellectus archetypus* coordinating the principles of their organization.[10] In mechanistic causality a cause produces an effect, which in turn might itself cause another effect, and so on in a serial train. However, in the biological realm, say, in the epigenesis of the fetus as described by Blumenbach (whom Kant knew and read), the various early stages make sense only in relation to their final product: that is, we have to conceive the final stage of development, which is an effect of the earlier stages, as if it were also the cause of the earlier stages. Indeed, all the causal sequences exhibited by organisms display a coherence that our reflection requires us to regard as if it were produced by an ideal plan realized in the whole. In this respect our reflective judgments about biological organization are directly analogous to our appreciation of artistic creations. For in judgments of beauty, we must also comprehend the arrangement of elements as the product of a harmonious ideal. Kant's theory of reflective judgment thus furnished the rationale for romantic naturalists, like Goethe, Schelling, Oken, and Coleridge, to insist on an aesthetic approach to biological science.

Kant nonetheless found rather seductive certain mechanistic proposals to explain the purposiveness manifested by animals. For instance, "the inclusion of so many animal genera in a common scheme . . . in terms of which a shortening of one part and an extension of another, a development of this part and a reduction of that, would thus be able to produce a great diversity of species"—such possibilities sparked a glimmer of hope that mechanistic interpretations of animal form might suffice.[11] Even a "daring adventure of reason," prompted by

10. Immanuel Kant, *Kritik der Urteilskraft,* A346–47, B350–51, in *Immanuel Kant Werke,* ed. Wilhelm Weischedel (Wiesbaden: Insel-Verlag, 1957), 5: 526.

11. Ibid., A363–64, B638 (Weischedel, p. 538). In Kant's suggestion here one can recognize the seeds of various theories of the archetype, as pursued by Goethe, Oken, von Baer, Green, and Owen (see below in this chapter and in the next for elaboration).

the familial resemblances of different species, might suggest that mother earth, through spontaneous generation, could in the beginning "have given birth to creatures of a less purposive form, and these likewise might have structured themselves in greater measure to their geographical circumstances and to their relations with each other," until finally the present species would have emerged.[12] Yet in the Kantian epistemology, these developmental hypotheses concerning species, aside from the lack of any real evidence supporting them, could not escape the antinomy of deriving purpose from mechanism. For the human mind to even entertain such hypotheses, it had to relate them to an ideal plan, an archetype, as the ground for their possibility.

According to Kant's critical idealism, neither efficient causal relationships nor teleological ordering were supplied by raw sensation; these rather flowed from the subjective side, through the activities of the ego that transcended experience of objects and its phenomenal self. Thus human experience became possible only when, within the ego's unity of apprehension, the categories of cause, substance, unity, and the rest prescribed by the Kantian logic organized sensations into intelligible, objective structures and when, in comprehending the nature of living creatures, additional teleological assumptions rendered such structures biologically meaningful. Kant, however, denied to the teleological functions of judgment the same logical status that the categories of understanding had in the constitution of scientifically objective experience. In the Kantian epistemology, ideas of purpose could only suggestively guide the scientific zoologist in reductively analyzing organisms into their determinate causal relations. Yet the logical need to comprehend organisms as more than causal mechanisms under the efficient control of the environment, the intellectual requirement of perceiving them as if their individual parts were rationally ordered to serve overall ends—these subjective epistemologi-

12. Ibid., A365, B369–70 (Weischedel, p. 539).

cal demands indicated a residue that would not give up its intelligibility to deterministic, Newtonian science. By the very structure of our judgment we had to regard the morphology of creatures and their systematic species relations "as if [*als ob*]" they had been purposefully designed, ultimately by an intelligent Creator, according to an ideal archetype; yet simultaneously we had to understand that such supposition had no objective ground, either in the necessary categorization of experience or, as far as we could ever know, in the external world on the other side of experience. In Kant's view, the phenomenal world of nature, interlaced with determinate causal relations and subjectively tinged with purpose, did not necessarily reflect the ultimately real, noumenal world, that hidden world of the "thing-in-itself," where God proposed and the transcendental ego disposed. The reality that went beyond experience had to remain unknown to scientific understanding, though, according to Kant, it could be approached through moral judgment.

Kant's young disciple Johann Gottlieb Fichte (1762–1814) rejected the assumption of an objective noumenal reality that supplied raw material for the noumenal ego to organize. According to Fichte, all of reality derived from the subjective ego: the ultimately affirming transcendental ego posits both the limited, empirical ego and the natural world. In Fichte's absolute idealism the Kantian noumenal object fell back into the maw of the pure, transcendental ego. The reality of biological organisms and their relationships thus would depend on the dynamic and nondeterministic activity of their subjective monopole. Fichte, initially a devoted disciple of Kant, was in turn adored by Schelling, who, as every good pupil must, came to sympathetic dissent.

Friedrich Schelling (1775–1854) was recommended for a professorship at Jena by Goethe; and there in 1798, at the remarkable age of twenty-three, he became Fichte's close colleague. Initially he began with Fichte's assumption that the empirical ego and the natural world flowed entirely from the infinite, subjective ego. In the first edition (1897) of his *Ideen zu einer Philosophie der Natur,* he, like Fichte, attempted to divest the objective noumenal thing-in-itself of any power, to reduce not only all necessary features of experience of nature to activity of the transcendental ego, as Kant had, but to reduce the contingent aspects of nature also to this subjective source. "Thus all representations of an external world," he concluded in the *Ideen,* "must develop out of myself."[13]

Just this reduction of all contingency to the subject struck Goethe as going too far. He complained to his friend, the poet Friedrich Schiller, that the young idealist failed to assess adequately our unanticipated experience of the natural world. We stand before the internal purposive structure of organisms just as the astonished Chinese stood before the German chafing dish.[14] But Goethe seems not to have quite grasped the radical character of Schelling's idealism. According to his Fichtean doctrine, the finite self, whence the representations of nature derived, only opened a window through which the infinite ego acted. Hence, the resistance and constraint felt when we comprehended nature, the recognition of necessity in our ordinary commerce with the world—these had a deeper source than merely the individual self. This feeling of necessity in nature could also be recognized in our perceptions of purposiveness, certainly a phenomenological appreciation that should have comforted Goethe. Moreover, the goal-directedness of organisms now had a proper metaphysical foundation: intelligent design in the world could be explained as a product of ultimate subjectivity, which, insofar

13. Friedrich Wilhelm von Schelling, *Ideen zu einer Philosophie der Natur,* in *Sämmtliche Werke,* ed. K. F. A. Schelling (Stuttgart and Augsburg: Cotta'scher Verlag, 1856–61), 2: 33–34.

14. Johann Wolfgang von Goethe to Friedrich Schiller (6 January 1798), *Goethe, Die Schriften zur Naturwissenschaft,* 2d division, vol. 9b: *Zur Morphologie, von 1796 bis 1815,* ed. Dorothea Kuhn (Weimar: Böhlaus Nachfolger, 1986), pp. 128–29. Goethe had taken Schelling's newly published *Ideen* on his trip to Switzerland. He reported that it offered his company much conversation and that he was anxious to pursue it in detail with Schiller on his return.

as it transcended individual consciousness, was also ultimate objectivity. Because of this subjective foundation for nature, "in early times human thought was led to the idea of self-organizing material, and, since organization is comprehensible only in relation to a mind [*Geist*], to an original union of mind and matter in these natural objects."[15] Schelling regarded this primitively original, creative and aesthetic appreciation of living nature as coming much closer to the truth—a Goethean truth—than ever could be achieved by the analytic techniques of reflective philosophy, which by separating mind from nature delivered up "a dead object."[16]

Shortly after the publication of the *Ideen,* and perhaps as a result of his close association with Fichte, or even because of rumors of Goethe's criticism, Schelling began to slide to the other side of the Kantian balance, moving away from the over-weighted subjective idealism of his Jena colleague. His new love was Spinoza. In the second edition (1803) of the *Ideen,* he urged acceptance of the principal Spinozistic insight that "the absolute-ideal is also the absolute-real."[17] He now elaborated his *Identitätsphilosophie,* which held that the primal infinite reality had from the very beginning both mental and material features. In a kind of Neoplatonic emanation, the phenomenal world of nature flowed from the infinite reality, just as the ideal world of mind emerged from this same absolute being. In an ascending dialectic, nature progressed gradually through higher stages, with each stage embodying a more complete ideal type (*Vorbild*): "so reason is symbolized in the organism just as the absolute knowledge-act is symbolized in eternal nature."[18] Our scientific knowledge moved in parallel, coming to a kind of finite perfection, so Schelling modestly estimated, in transcendental philosophy. The telos for this dual progressive development was, on one side, greater individuality in nature—as it advanced from less indi-

15. Schelling, *Ideen zu einer Philosophie der Natur,* p. 47.
16. Ibid., p. 48.
17. Ibid., p. 58.
18. Ibid., p. 69.

viduated, less complex animals to more complex and perfectly individuated creatures, reaching its apotheosis in man. And this natural evolution would be matched by the advance of human reason, until these finite strands of nature and mind were rewoven into the absolute.

The new pattern of objectivism in Schelling's thought appeared gradually during the few years on either side of 1800. In the *Erster Entwurf eines Systems der Naturphilosophie* (1799), he described the progressive development of nature as a "dynamic evolution [*dynamische Evolution*]."[19] According to this theory, the original seed, which gave rise to the first individuals in a species, would imperfectly embody the species concept (*Gattungsbegriff*). But as the direction of the *Bildungstrieb* became determined by shaping pressures of the external environment, later individuals would express more and more species features—so that with the many races evolving, nature would come more completely to realize in the varieties of individuals the full concept of the species.

Schelling adopted the term *Evolution* from Haller's embryology. He added the adjective *dynamische* in light of the criticisms of Wolff and Blumenbach. He agreed with these zoologists that fetal evolution occurred epigenetically, urged by the *Bildungstrieb* to advance through higher developmental stages (*Entwicklungstufe*).[20] To these borrowed embryological ideas Schelling applied the hermeneutic apparatus of the *Naturphilosophie:* the dynamic evolution of the embryo furnished a phenomenal sign of the deeper logical development of absolute being through its evolutionary stages. The microcosm of the animal womb recapitulated in figures the macrocosm of the womb of nature. It would be a mistake, however, to gloss Schelling's conception of recapitulation in more sober terms of historical species transformation, such as Erasmus Darwin proposed at this time in his *Zoonomia,* a

19. Friedrich Wilhelm von Schelling, *Erster Entwurf eines Systems der Naturphilosophie* (1799), in *Sämtliche Werke,* 3: 61, 206, 292.

20. Ibid., pp. 59–61.

work with which Schelling was quite familiar.[21] Schelling himself cautioned that "the assertion that the different [species] organizations have actually been formed through a gradual development of one out of the other is a misunderstanding of the idea, which really lies in reason."[22] He allowed for an evolution of varieties within species but insisted that the essential forms of species only evolved in a logical, atemporal way, moving through the deeper dialectic of being.

Kant and Schelling formulated the intellectual arguments for *Naturphilosophie,* but its soul flowed from the pen of Johann Wolfgang von Goethe (1749–1832). Through the nineteenth century Goethe became the central icon not only of German literary culture but of German science as well, certainly biological science. Carl Gustav Carus (1789–1869), whose *Lehrbuch der Zootomie* (1818) domesticated some of the wilder impulses of transcendental morphology,[23] loved to cite Goethe's approbation of his work—even if the great poet approved Carus's views because he believed they reflected his own ideas.[24] Hermann von Helmholtz (1821–94), nineteenth-century Germany's most renowned scientist, assured the audience for his popular lecture series that Goethe's zoological theories, especially his theory of type, had been confirmed by subsequent researchers and that even his theory of colors had gone to magnificent ruin on the rocks of his poetic genius. Ernst Haeckel provides a final example of the depth of Goethe's scientific impact. Each of the thirty chapters of Haeckel's great two-volume *Generelle Morphologie der Organismen* (1866) begins with some passage from Goethe that represents an aesthetic understanding

21. Ibid., p. 167.

22. Ibid., p. 63.

23. See nn. 8 and 77 for a sketch of some of Carus's basic notions.

24. Carus cited a letter of praise from Goethe (1818) both in the 2d ed. of his *Lehrbuch* and in his autobiography. See Carl Gustav Carus, *Lehrbuch der vergleichenden Zootomie,* 2d ed. (Leipzig: Fleischer, 1834), 1: viii; and *Lebenserinnerungen und Denkwürdigkeiten* (Leipzig: Brockhaus, 1865), 1: 195–96.

of nature's developmental processes, as, for instance, these
lines from the "Metamorphosis of Animals" that introduce the
seventeenth chapter:

> Every animal an end in itself springs forth perfect
> From the womb of nature and produces perfect
>   children.
> All the limbs form themselves according to eternal
>   laws,
> And the rarest forms preserve in secret the original
>   pattern [*Urbild*].
> So each foot, long or short, moves in complete
>   harmony
> With the essence of the animal and its
>   requirements.
> For all the animate limbs never contradict
>   themselves,
> Rather all are directed toward life.
> The form of the animal determines its way of life,
> While its ways powerfully react on every form,
> And so the ordered formation [*Bildung*] holds fast,
> Though inclining to change through the action of
>   external conditions.[25]

Haeckel's rhetorical device was appropriate, since he regarded Goethe as a founder, along with Lamarck and Darwin,
of evolutionary theory—or at least the theory he wished to
advance. This stanza from Goethe's lyric, however, suggests
that his version of evolution would not excite a tingle of recognition in a modern biologist, though I think Darwin would
have appreciated it. Haeckel was inspired by it. Goethe's
vision of the developmental process was distinctly morphological, as was Haeckel's own.

25. Johann Wolfgang von Goethe, "Metamorphose der Thiere," *Goethes
Gedichte in Zeitlicher Folge,* ed. Heinz Nicolai, 2d ed. (Frankfurt am Main:
Insel Verlag, 1982), pp. 565–66. The poem was written in Weimar, sometime
between 1798 and 1799. It opens chap. 17 of Haeckel's *Generelle Morphologie der Organismen,* 2 vols. (Berlin: Reimer, 1866), 2: 2. I have truncated the
stanza a bit.

Goethe's interest in artistic representation initially motivated his anatomical studies. In 1781 he took time from his court duties at Weimer to study anatomy at Jena, which would become the central command post for *Naturphilosophie*.[26] His comparative investigations through the 1780s supported his growing belief that a single plan underlay the human and animal form, a belief crystallized in his discovery of the intermaxillary bone in the human skull. The anatomist Peter Camper (1722–89), following a long tradition beginning with Vesalius, had distinguished man from the apes because of the apparent absence of this beastly bone (see fig. 4). Goethe believed the structure, which in most vertebrates lies lateral and inferior to the nasal cavity—he found a nice example in the walrus—had also to be present in man if nature were a harmonious whole (see fig. 5). He extended his search to the human fetus, where he found the bone in a separate condition before it fused with the upper jaw. He reported his discovery to a friend in March of 1784.[27] Though Camper, Blumenbach, and others initially disputed the presence of the intermaxillary in man, zoologists just after the turn of the century treated as commonplace its appearance in the human fetus.[28] This discovery convinced Goethe of the crucial importance of em-

26. Fichte and Schelling taught at Jena at the beginning of the century, and Haeckel dominated the university from the 1860s until his death in 1919.

27. The date is documented in *Goethe, Die Schriften zur Naturwissenschaft,* 2d division, vol. 9a: *Zur Morphologie: von den anfängen bis 1795, ergänzungen und erläuterungen,* ed. Dorothea Kuhn (Weimar: Böhlaus Nachfolger, 1977), p. 476. The pamphlet he wrote in 1786 and circulated to friends, "Dem Menschen wie den Tieren ist ein Zwischenknochen der obern Kinnlade zuzuschreiben," was first published in 1820 in the second of six collections of his *Zur Morphologie* (1817–24). See Johann Wolfgang von Goethe, *Goethe, Die Schriften zur Naturwissenschaft,* 1st division, vol. 9: *Morphologische Hefte,* ed. Dorothea Kuhn (Weimar: Böhlaus Nachfolger, 1954), pp. 154–86. Goethe was happy to learn that Félix Vicq d'Azyr had already described the human-fetal intermaxilliary in 1780, though he neglected to mention Vicq's anticipation in his own later publication of the discovery. See *Goethe, Die Schriften zur Naturwissenschaft,* 2d division, 9a: 482.

28. See, for example, Carus, *Zootomie,* 1st ed., pp. 172–73.

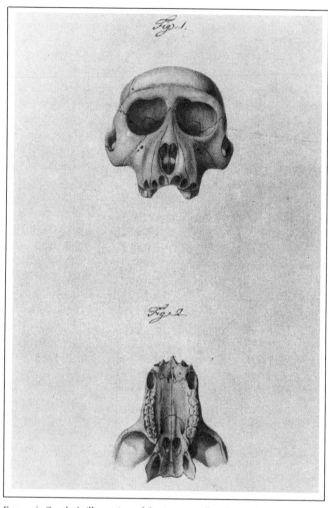

FIGURE 4. Goethe's illustration of the intermaxillary bone (lying lateral and inferior to the nasal cavity) in an ape.

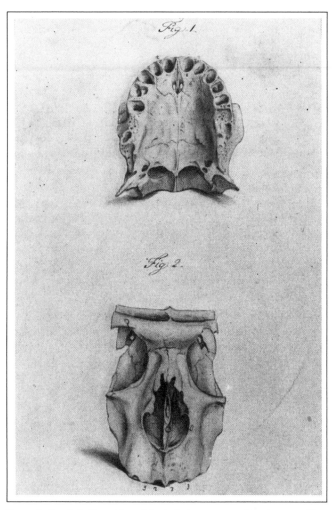

FIGURE 5. Goethe's illustration of the indistinct sutures of the intermaxillary bone in man.

bryological study for establishing the internal pattern of organisms and for dispelling the belief "that what is hidden in the species or in the individual must thus be absent. By this means we learn to see with the eyes of the mind, without which we must stumble around blindly, especially when researching into nature."[29]

Goethe's most famous morphological study was in botany, his *Die Metamorphose der Pflanzen,* which appeared in 1790. In that brief treatise, he argued in a way congenial to the teleological theory of Kant's *Kritik der Urteilskraft* (1790), a harmony of conception he shortly thereafter realized. From his somewhat limited comparative and developmental investigations, Goethe concluded that the different parts of plants could be understood as transformations of one underlying structure, that of the ideal leaf. But the convergence of Goethe's ideas on those of Kant, especially after he absorbed the Third Critique, appears most obvious in his extensive and detailed comparative studies of animal osteology.

Goethe began with the metaphysical conviction that the vertebrates were "all formed according to a single, original plan [*Urbild*]."[30] But he justified this belief methodologically and experientially. The very concept of the type, he argued, required us not to identify it with any particular species or genus, since the type must be common to the whole vertebrate realm. Typological thinking, he urged, was methodologically required for the comparative zoologist, since (as Plato knew) the various species could only be descriptively related to one another through a model that lacked their individual differences.[31] The construction of such a model must be governed, Goethe insisted, by comparative and embryo-

29. Johann Wolfgang von Goethe, "Erster Entwurf einer allgemeinen Einleitung in die vergleichende Anatomie, ausgehend von der Osteologie" (1795), *Morphologische Hefte,* p. 138.

30. Johann Wolfgang von Goethe, "Vorträge über die drei ersten Kapitel des Entwurfs einer allgemeinen Einleitung in die vergleichende Anatomie" (1796), *Morphologische Hefte,* p. 198.

31. Goethe, "Erster Entwurf einer allegemeinen Einleitung in die vergleichende Anatomie, ausgehend von der Osteologie," p. 121.

logical discovery.[32] From experimental analysis, the investigator would be able to abstract the *Urbild,* or, as he quoted from Kant, "the *intellectus archetypus*" of the vertebrate skeleton.[33] The archetype, then, would consist of an abstract pattern of bones. Such pattern would be identifiable in individual animals, since comparable structures would retain their same relative positions throughout the vertebrates, but would be differently elaborated depending on the species—thus the bones in the legs of man and steer would exhibit the same abstract, topological relationships, while differing greatly in detail.[34] According to Goethe's theory of the archetype, individual bones, both in their structural character and functional operations, formed a coherent whole, an internal purposiveness, which from a casual external perspective might seem superfluous or useless.

> We think of the isolated animal as a small universe, which acts and exists for itself. Accordingly, each animal is an end in itself; and because all of its parts stand in direct interaction, because they have a relation with each other and are renewed in the cycle of life, each animal can be looked upon as physiologically complete. No part of the animal, seen from the inside, is useless, or, as one sometimes imagines, arbitrarily produced by the formative drive [*Bildungstrieb*]; although, parts seen externally can seem useless, since the inner coherence of animal nature molded them without regard to external relationships.[35]

32. Goethe, "Vorträge über die drei ersten Kapitel des Entwurfs einer allgemeinen Einleitung in die vergleichende Anatomie," p. 200.

33. Johann Wolfgang von Goethe, "Anschauende Urteilskraft," *Morphologische Hefte,* pp. 95–96. Goethe hoped that further study of the invertebrates would enable researchers to generalize the "Urtier," or archetypal animal, to include those creatures. See "Erster Entwurf einer allgemeinen Einleitung in die vergleichende Anatomie," pp. 122–24.

34. Goethe, "Erster Entwurf einer allgemeinen Einleitung in die vergleichende Anatomie," p. 134.

35. Ibid., p. 125. The theory of the animal archetype constituted one part of Goethe's larger morphology, or doctrine of forms in nature. In his *Zur Mor-*

Goethe ignored the larger teleological perspective of an animal's adaptation to its environment; he rather focused on the internal adjustment of organic part to organic part. At times he suggested that the metamorphosis of parts responded to a Blumenbachean *Bildungstrieb*;[36] at other times, separated by the breath of a few sentences, he ascribed the transformation of parts, much like Buffon, to the impingements of external conditions in an animal's environment.[37] Whatever his exact belief about the forces altering animal organs, he consistently held that the archetypal pattern constrained the possibilities of metamorphosis.

Goethe's theory of the archetype had another important feature that through intellectual diffusion became essential to later English versions of the archetype; this was the idea that the various bones of the vertebrates were, like the parts of the plant, but repetitions of the same bone variously metamorphosized. The vertebrate archetype, in Goethe's view, consisted of a transformation and repetition of the vertebra itself. He came to this view partly on the basis of beliefs about the unitary structure of plants, but also because of his conviction that the bones of the skull were really six transformed vertebrae.

---

phologie, Goethe defined morphology in this way: "Morphology rests on the conviction that everything which exists must signify and reveal itself. From the first physical and chemical elements to the mental expression of man we find this fundamental principle to hold. We turn immediately to that which has form. The inorganic, the vegetative, the animal, the human—each signifies itself, each appears as what it is to our external and our internal sense. Form is changeable, becoming, passing. The doctrine of form is the doctrine of alteration. The doctrine of metamorphosis is the key to all the signs of nature." See *Goethe, Die Schriften zur Naturwissenschaft,* 1st division, 10: 128. Goethe's morphology, as indicated in this passage, was dynamic and extended to all the structural changes exhibited in nature. Karl Friedrich Burdach first used the term *Morphologie* in print in 1800, but he confined his usage to the investigation of the human form in the context of medicine. Goethe had already coined the term in correspondence with Schiller in 1796. See *Goethe, Die Schriften zur Naturwissenschaft,* 2d division, 9b: 418.

36. Ibid., p. 125.
37. Ibid., p. 124.

Goethe first published his vertebral theory of the skull in 1820, in the second number of his *Zur Morphologie.*[38] Some distilled vitriol etched the brief announcement of his long-held belief; for Oken had claimed the idea for himself in 1807, and further argued his claim against all rivals in 1817.[39] Goethe maintained, and his letter to Karoline Herder confirms his story, that the hypothesis came to him in 1790 while he was sporting with an animal skull in a Jewish cemetery in Venice.[40] Since he subsequently discussed his view among many naturalists, he thought Oken had gotten wind of it, enough to inflate his rival's sails. In any case, the idea that the complex formation of the vertebrate skull could be explained as the metamorphosis and repetition of a single part suggested to Goethe that the vertebrate archetype revealed not only the unity hidden beneath diverse transformations of species (what Owen would later call "special homology") but itself harbored a yet deeper and more profound unity of part to whole (Owen's "general homology").

Haeckel identified Goethe as a principal founder of evolutionary theory. Goethe was not, to be sure, an evolutionist in the manner of Erasmus Darwin or Jean-Baptiste de Lamarck, both of whom he read, but in the manner of Schelling. Schelling discovered in Spinoza an objective ground for the dynamic evolution of nature's purposiveness, but Goethe from his youth had already been led by the Jewish lens grinder to worship *Deus sive Natura.*[41] Thus for Goethe, as for Schelling, embryonic development more symbolized the deeper emanations of infinite being rather than revealed a distinct historical process of species transformations. Goethe's more settled

38. This was appended to his essay on the intermaxillary bone in man. See Goethe, *Morphologische Hefte,* p. 185.

39. For bibliographic information, see n. 45, below.

40. See Goethe's letter to Karoline Herder (4 May 1790) in *Goethe: Die Schriften zur Naturwissenschaft,* 2d division, 9a: 392.

41. Johann Wolfgang von Goethe, *Aus meinen Leben: Dichtung und Wahrheit* (1811–33), 2d ed. (Berlin und Weimar: Aufbau Verlag, 1984), p. 623: "This mind that so decisively affected me and has so greatly influenced my whole way of thought was Spinoza."

attitude anchors the sentiments of a quatrain included in *Zur Morphologie:*

> Natura infinita est,
> sed qui symbola animadverterit
> omnia intelliget
> licet non omnino.[42]

Kant, Schelling, and Goethe absorbed many theories advanced by zoologists during the last part of the eighteenth century. But then, by a kind of creative dialectic, these distinctively German thinkers sparked to life the program of transcendental morphology. The biologists whom we have historically denominated transcendental morphologists certainly differed among themselves on various issues; but they were united through many of the ideas given inspired formulation by Kant, Schelling, and Goethe. So, for instance, the biologists discussed in the remaining sections of this chapter would agree that a unity of law or reason bound together the parts and whole of the natural world. Such reflectively faceted reason showed forth in various ways: some researchers insisted on the parallel relations of fetal development and species hierarchy, while others would find each animal part but a type of the fundamental unit, as in the case of Oken's and Owen's theories of the vertebral skull. A virtually defining feature of transcendental morphology was the belief that the animal world could be ordered according to a few ideal types, types which expressed an inner purposiveness. The type, as generally conceived, was an abstract pattern that consisted of a topologically constant arrangement of parts, which themselves were transformations of a single element. These archetypes ruled natural development. So even von Baer, the least excessive zoologist of the period, argued in a Schellingesque fashion, as we will see in a moment, that the essence or idea of the type causally directed embryogenesis. The transcendental

42. Johann Wolfgang von Goethe, "Zwischenrede," *Morphologische Hefte,* p. 89: "Nature is infinite, but he who would attend to symbols will understand everything, though not quite."

morphologists had little difficulty in arguing that the ideal determined the disposition of the real, since they implicitly, sometimes explicitly, adopted the Spinoza-Schelling metaphysics, which identified the ideal and the materially real. The attractive force that joined all of these thinkers, from Oken and von Baer in Germany to Owen and Chambers in England—and, I will argue, Darwin as well—was, in Schelling's terms, *dynamische Evolution,* dressed out, as it would be in the various semantic guises of development, *Entwickelung,* metamorphosis, transmutation, and the rest. The concept of dynamic evolution implied that organisms were impelled— either in the deeper, atemporal mode of the Spinozistic *nisus,* or in the way of historical transformations—to greater perfection, though preserving all the while a rational ideal. In all of these respects, the prototype of the transcendental morphologist, if not the perfection of the type, was Lorenz Oken (1779–1851).

Oken's Transcendental Morphology

Oken's ideas came to be regarded as the epitome of German romantic biology. His transcendental vision of nature, which from our perspective seems to cloud over his comprehensive fund of nineteenth-century knowledge, enthralled a whole generation of biologists, even those of a more stolid scientific constitution, such as Richard Owen, who adopted his theory of the vertebral skull, and Louis Agassiz, who studied with him at Munich. The romantic strains of his approach reverberate in the very opening lines of a lecture series he gave in 1805 as *Privatdozent* in Göttingen:

> What is the animal kingdom other than an anatomized man, the macrocosm of the microcosm? In the former there lies open and expanded in the most beautiful order what in the latter is collected into small organs, though arranged in that same beautiful way. As the bloom takes up lovingly and intimately into itself all the parts of the plant, and so elevated in shimmering arrayment offers them in sacrifice to

Phoebus and the eternally enhancing goddess of
life, so man spiritualizes all of those natures that stir
enclosed so miserably in the lowest fleshy vesicles
and manifests a brilliant resurrection of them in
himself.[43]

In *Die Zeugung,* his first major publication after his docto-
rate at Freiburg in 1804, Oken declared in his *Naturphilosophisch* fashion that the animal kingdom itself was a complete
animal, with each of the component animals, from the infu-
sorium up, supplying the graded hierarchy of organs. Within
the whole and the parts parallels abounded. For instance,
"the [mammalian] fetus, through the course of the several
forms of its existence, is the whole animal": it sequentially
acquires organs which characterize the hierarchy of the kingdom of life, passing through stages analogous to that of polyp,
plant, insect, snail, fish, and amphibian, finally settling into
the class to which it belongs.[44] Two years later he proposed
that particular parallel Richard Owen would later embrace
(much, Huxley thought, like an iron maiden): namely, the
repetition of modified vertebrae to form segments of the
skull. He quickly extended this theory to argue—in the
spirit if not the letter of Kant's and Goethe's notions of the
archetype—that "the entire man is only a vertebra."[45] In

43. Lorenz Oken, *Abriss des Systems der Biologie* (Göttingen: Vandenhoek
und Ruprecht, 1805), p. iii. The natures enclosed in the lowest fleshy vesicles
are infusoria, or uranimals (*Urthiere*), into which he believed the flesh of
higher animals could be resolved.

44. Lorenz Oken, *Die Zeugung* (Bamberg und Wirzburg: Goebhardt,
1805), pp. 146–47.

45. Lorenz Oken, *Ueber die Bedeutung der Schädelknochen* (Bamberg:
Göbhardt, 1807), p.1. Oken's theory of the vertebrate skull, the subject of his
inaugural dissertation at Jena, held that the various bones of the skull were
only elaborations of three vertebrae, named by him the "ear vertebra," the
"jawbone vertebra," and the "eye vertebra," according to the prominent skull
parts to which they gave rise (p.6). See Oken's more extensive, later discussion of his theory in his own review of "Ueber die Bedeutung der
Schädelknochen," *Isis oder Encylopädische Zeitung* 1 (1817): 1204–9. He
later expanded the number to four constituent vertebrae, since the bones of

FIGURE 6. Lorenz Oken, 1779–1851, an etching done for his friends. Well-come Institute Library, London.

that great monument to transcendental morphology, the *Lehrbuch der Naturphilosophie* (1810), Oken uncovered parallels without end. For example, just as semen gives rise to the polyplike embryo in sexual generation, so sea slime produces infusoria through equivocal (spontaneous) generation. Thus "man too is a child of the warm and shallow places of the sea."[46] From this *Naturphilosophisch* perspective, Oken concluded that "animals are only the fetal stages of man"[47]— a maxim, in Serres's version, that Darwin would later scribble into one of his transmutation notebooks.

Oken's *Naturphilosophie* was not a transmutationist tract. As with Schelling, his comparisons remained ideal. The mammalian fetus evolved (*entwickelte*) through stages of organ development that paralleled the hierarchy of the animal kingdom, insofar as each animal class represented particular organs: infusoria were but digestive tubes, stomachs; bivalves first displayed a vascular system, liver, and ovaries; snails exhibited (following a line that would have delighted Freud) heart, testicles, and penis. Several authors writing at the same time as Oken, however, seized upon the possibility of a temporal parallel, the idea that the embryos of higher animals advanced through the same stages that their ancestors had actually reached in their progressive transformation over generations.

Evolutionary Recapitulationism of Tiedemann, Treviranus, and Meckel

The temporal parallel certainly struck Friedrich Tiedemann (1781–1861), who economically and daringly encompassed

---

the body were the armor of the senses and the skull encased four sense organs (touch being protected by elaborations of the rest of the vertebrae). See his *Allgemeine Naturgeschichte für alle Stände,* vol. 4 (Stuttgart: Hoffmann, 1833), p. 167. Despite Goethe's suspicions, Oken seems to have worked out his theory of the skull independently of the great poet.

46. Lorenz Oken, *Lehrbuch der Naturphilosophie,* 2d ed. (Jena: Frommann, 1831), pp. 156, 148.

47. Ibid., p. 388.

the most recently formulated biological theory in his uncompleted three-volume *Zoologie* (1808–14). He seems to
have balanced the romantic idealism of Schelling, his teacher
at Würzburg, against the sober instruction he received in
Paris with Georges Cuvier. Undoubtedly Cuvier thought
Tiedemann's views tinged with everything but sobriety. For
Tiedemann introduced recapitulation theory as patent:
"Every animal, until it reaches its own structure, passes
through the organization of one or more of the animal classes
standing under it, which is to say, every animal begins its
metamorphosis with the simplest animal organization."[48]
Tiedemann made comparisons—for example, at five months
the eye of the human embryo is similar to that of a fish[49]—but
they were hardly numerous or systematic. And though he
claimed to have undertaken a great many dissections of animals, his embryology appears to owe as much to Johann
Friedrich Meckel's early *Abhandlungen aus der menschlichen
und vergleichenden Anatomie und Physiologie* (1806),
which Meckel wrote while he and Tiedemann were under
Cuvier's tutelage. Only later did he undertake those detailed
comparative studies, especially of recapitulated structures of
the human fetal brain, that brought him to the attention of
English and French anatomists.[50] Though short on actual
demonstrative comparisons in the *Zoologie,* Tiedemann frequently reiterated the general principle of recapitulation,
and, more significantly, made the temporal analogy good in
application to the animal kingdom. "Just as each individual,"
he claimed, "begins with the simplest formation and during

48. Friedrich Tiedemann, *Zoologie, zu seinen Vorlesungen Entworfen,*
vol. 1 (Landshut: Weber, 1808), pp. 56–57.

49. Ibid., p. 173.

50. The work of Tiedemann familiar, at least in translation, to Charles Lyell
and other British naturalists was *Anatomie und Bildungsgeschichte des
Gehirns im Foetus des Menschen nebst einer vergleichenden Darstellung des
Hirnbaues in den Thieren* (Nuremberg: Steineschen Buchhandlung, 1816).
The English translation appeared in 1826, done out of the French language
version.

FIGURE 7. Friedrich Tiedemann, 1781–1861, early German recapitulationist and evolutionist; engraving done in 1830 from a portrait. Wellcome Institute Library, London.

its metamorphosis becomes more evolved [*entwickelt*] and developed, so the entire animal organism [i.e., kingdom] seems to have begun its evolution [*Entwickelung*] with the simplest animal forms, that is with the animals of the lowest classes."[51]

Tiedemann backed his claim of an actual transformation of species by appealing to the arguments of Gottfried Reinhold Treviranus (1776–1837), who maintained, in the third volume (1805) of his *Biologie, oder Philosophie der lebenden Natur* (1802–22), that the progressive deposition of fossil remains, with those of the simplest and no longer living animals at the stratigraphically lower layers, indicated an actual transformation of species in time.[52] Treviranus himself suggested, in the spirit of Kielmeyer, that species development followed the same laws as individual development;[53] but generally this idea of parallel evolution weighed lightly in his work. Of more significance in tipping his opinion to species transformation appears to have been his huge compilation of paleontological evidence, along with the *Naturphilosophish* idea of progressive alteration. It was left to Tiedemann to join firmly the evidence for species transformation with the developmental analyses of comparative embryology. And this he did by forcefully urging that the gradation of fossil deposits paralleled the developmental stages of the individual organism:

> It is clear from the previous propositions that from the oldest strata of the earth to the most recent, there appears a graduated series of fossil remains, from the most simply organized animals, the polyps, to the most complex, the mammals. It is evident too that the entire animal kingdom has its evolutionary periods [*Entwickelungsperioden*], similar to the periods which are expressed in individual organisms. Those animal species and genera which have under-

51. Tiedemann, *Zoologie*, 1: 64–65.

52. Gottfried Reinhold Treviranus, *Biologie, oder Philosophie der lebenden Natur,* vol. 3 (Göttingen: Röwer, 1805), pp. 3–226.

53. Ibid., pp. 40, 225–26.

gone an evolution [*Entwickelung*] can be compared
with the organs that in the course of the evolution of
each animal have vanished—for example, after birth
the vascular system of the navel, the thymus gland,
etc. have disappeared in human beings; with frogs,
the tail finally vanishes. Just as these parts, the organs
in the evolutionary periods of the individual organism, have vanished, so have animals, the organisms
of the evolutionary periods of the animal kingdom.[54]

It may have been just this passage that suggested to von Baer
the conceptual—and historical—relation of the idea of recapitulation to that of species change.

In his *Entwickelungsgeschichte der Thiere,* von Baer isolated as the source for the theory of species transmutation the
early recapitulation thesis—the belief that higher animals, in
their embryological development, passed through the forms
of *now-existing* lower animals.[55] He complained, much like
Schelling, that even if this early version of recapitulation were
true, naturalists tended to forget its metaphorical features:

One gradually learned to think of the different animal forms as evolving [*entwickelt sich*] out of one
another—and then shortly to forget that this metamorphosis was only a mode of conception. Fortified
by the fact that in the oldest layers of the earth no remains from vertebrates were to be found, naturalists
believed they could prove that such unfolding of the
different animal forms was historically grounded.
They then related with complete seriousness and in
detail how such forms arose from one another. Noth-

54. Tiedemann, *Zoologie,* 1: 73–74.

55. Von Baer was quite aware that the principle of recapitulation, even in its
older version, did not antedate several vague notions of species transformation (*Entwickelungsgeschichte der Thiere,* p. 201 n). Lamarck's theory, for instance, seems to owe nothing to the principle. However, von Baer did suggest
that recapitulation theory became fused with and *logically* preceded that of
species transformation. Certainly by his time, as I indicate immediately in the
text below, the conception of morphological transformation had become applied indifferently to embryo and species.

ing was easier. A fish that swam upon the land wished
to go for a walk, but could not use its fins. The fins
shrunk in breadth from want of exercise and grew in
length. This went on through generations for a cou-
ple of centuries. So it is no wonder that out of fins
feet have finally emerged.[56]

Tiedemann suggests and von Baer confirms the historical
transition that I have been charting in this chapter: the passage from embryological schemes of evolution to notions of
species evolution. Historically one strong current has led
from the old theory of evolution, i.e., the idea that the embryo
is a miniature adult of its species, through the first version of
recapitulation, i.e., the theory that the embryo of higher animals passes through stages represented by the adult forms of
now-existing lower animals, to the theory of species evolution, i.e., that adult forms of existing creatures unfolded from
forms of their *no-longer-existing,* lower ancestors. The new
principle of recapitulation—which holds that the embryo
of higher organisms evolves through forms of *extinct* creatures—this new principle became in von Baer's time the living descendent of these historical developments, and the
residual sign of connection between the old theory of embryological evolution and the modern theory of species evolution. Indeed, according to von Baer, the idea of evolution of
morphological structure in the individual and in the species
melted into the single notion of the transformation of anatomical pattern, whether in the embryo or in the lineage:
"There is only *one* way of metamorphosis, that of the further
development [*Ausbildung*] either achieved in one individual
(*the individual metamorphosis*), or through the different ani-

56. Von Baer, *Entwickelungsgeschichte der Thiere,* p. 200. Von Baer seems
to echo here the similar remarks of Georges Cuvier in his devastating eulogy
to his dead colleague Lamarck. Cuvier also thought that "according to
Lamarck, nothing is so easy to conceive" as the production of new organs
through use and disuse. See Georges Cuvier, "Eloge de M. de Lamarck,"
Memoires de l'Academie des sciences, 2d ser., 13 (1835): i–xxxi. Relevant passages are on pp. xix–xx.

mal forms (*the metamorphosis of the animal kingdom*)."[57]
The general principle of recapitulation, I believe, reveals an
important origin of Darwin's theory and gives it historical
depth beyond that usually conceived by historians.

Early German biology was stimulated to find in matter itself principles of goal directedness (compliments of Kant)
and progressive development (inspired by Schelling), but
this did not mean its practitioners slid quickly into the
Urschleim of romantic philosophy.[58] Autenrieth, Tiedemann,
and Treviranus certainly engaged in exacting empirical work
and scoured the literature of their predecessors with an eye
cautious to avoid, as Treviranus put it, "Mystik und Schwär-
merei."[59] And while Oken may have sunk rather deeply into
this latter, he could still be read by Georges Cuvier, Karl Ernst
von Baer, Johann Meckel, Louis Agassiz, Richard Owen,
William Carpenter, and Ernst Haeckel with profit, as well as
with schadenfreude. Yet none of the early German pro-
moters of the parallel principle had the conceptual power or,
more importantly, the extraordinary empirical technique of
Johann Friedrich Meckel. Von Baer later confessed in his
autobiography that it was principally Meckel's work he had
set out to destroy.[60]

When von Baer had initially made the acquaintance of
Meckel at Halle in 1817, the older man was "the most famous
anatomist of his time."[61] Fame was a family tradition. His

57. Von Baer, *Entwickelungsgeschichte der Thiere,* p. 201.

58. Timothy Lenoir has argued persuasively for the role of Kant in making
teleological considerations respectable for German biologists of the nineteenth century. I believe, however, that Lenoir neglects the powerful impact
of Schelling and idealist philosophy in transforming Kant's epistemological
position into a metaphysical one that identified the mental and the material.
See Lenoir's admirable *Strategy of Life* (Chicago: University of Chicago Press,
1989). See also n. 81, below.

59. Treviranus, *Biologie,* 3: 545.

60. Karl Ernst von Baer, *Autobiography of Dr. Karl Ernst von Baer,* ed. Jane
Oppenheimer; trans. H. Schneider (Canton, Mass.: Science History Publications, 1986), p. 312.

61. Ibid., p. 154.

grandfather, the first Johann Friedrich, had been a favored disciple of Haller and had taught anatomy at Halle; and his father, Phillip Friedrich, a teacher of Wolff, was so devoted to anatomical science that he arranged to have his own skeleton mounted in the university museum, posthumously of course. Johann Friedrich and his brother David followed in these footsteps. In addition to the familial influences, Meckel studied with Cuvier at Paris and translated his mentor's anatomical work into German. He also brought out a German translation of Wolff's doctoral dissertation, vaulting his father's student to greater prominence. Just as Richard Owen was renowned as the English Cuvier and Louis Agassiz as the American, Meckel became known as the German Cuvier, even though his philosophical flights into evolution would have made the other Cuviers quite nauseated.

While at Paris with Cuvier, Meckel attempted to discover the functions of the thymus, thyroid, and adrenal glands by comparing their sizes both in relation to other organs and at different times in the growth of several species of animals, including man. He concluded, for example, that the adrenals had sexual functions because their malformations usually accompanied irregularities in the sex organs.[62] Meckel described this extraordinary study in the first part of his *Abhandlungen aus der menschlichen und vergleichenden Anatomie und Physiologie* (1806). This intricate comparative analysis led him, in the second part of the work, to investigate more generally the evolutionary history (*Entwickelungsgeschichte*) of the human embryo. With some difficulty he acquired, through the offices of Cuvier, six human fetuses of various ages and undertook a detailed comparison of the relative sizes and situations of their organs through advancing developmental stages. He also compared these parts with those of lower animals at similar levels of development. Though these latter comparisons were not perfect (e.g., whale embryos have relatively larger facial parts and smaller

62. See Lenoir's felicitous account in *Strategy of Life*, pp. 58–59.

brains than human embryos at comparable stages), his opinion, he declared, was not far different from Kielmeyer's,
namely, that "the human fetus in its development [*Entwickelung*] indicates the stages at which lower animals have
remained throughout their entire lives."[63]

This directive idea became Meckel's guiding investigative
principle throughout his career at Halle. In his series "Con-
tributions to Comparative Anatomy," published from 1808 to
1812, he continued to use embryological comparisons to establish the functions of different organs. But he also began to
employ the principle of recapitulation more generally, as the
title of the first part of his contribution of 1811 suggests: "Outline of a Representation of the Parallel that Exists between the
Embryological Conditions of the Higher Animals and the Permanent Conditions of the Lower."[64] In this contribution he
investigated in some detail the embryological changes undergone by the heart and vascular system of mammals; he attempted to show how their earlier stages resembled what
could be found in frogs, salamanders, lizards, and various insects. He also compared among several species, in a fairly un-
systematic and sometimes (to our eyes) fanciful way, the
nervous system, sensory organs, intestinal tract, sexual
organs, urinary tract, and skeletal system. Through these discovered and invented parallels Meckel moved to a conception about the unity of the animal kingdom that, on the one
hand, made his work the archetype of German transcendental anatomy, and on the other, the most interesting causal an-
tecedent to that set of ideas traveling under the name of
"Darwinian evolutionary theory."

Georges Cuvier had proposed in 1812, much in the spirit
of Kant and Goethe, that all animals expressed one of four
basic plans (*embranchements*) in their organization, that of

63. Johann Friedrich Meckel, *Abhandlungen aus der menschlichen und
vergleichenden Anatomie und Physiologie* (Halle: Hemmerde und Schwetschke, 1806), p. 294.

64. Johann Friederick Meckel, *Beyträge zur vergleichenden Anatomie,* vol.
2, no. 1 (Leipzig: Reclam, 1811), pp. 1–60.

FIGURE 8. Johann Friedrich Meckel, 1724–1774, who made recapitulation theory central to his comparative anatomy; engraving. Wellcome Institute Library, London.

the radiata (e.g., jellyfish and starfish), mollusca (e.g., clams and octopuses), articulata (e.g., bees and lobsters), or vertebrata (e.g., fish and men).[65] While each of the *embranchements* displayed enough variability to accommodate the particular structures of species within them, no species could bridge these highest classes—for instance, the exoskeleton of articulata could not fit into the tightly organized and functionally economical structure of the vertebrata. This principle of correlation of parts, according to which only certain sets of traits could operate harmoniously together, Cuvier based ultimately on the functional principle of the conditions of existence—namely, that an animal was fitted for a certain kind of life, that it squeezed into its particular environment as a hand into a glove. Any fundamental alteration of parts among themselves, which would also disconnect the animal's bindings to its environment, could only extinguish life. These principles of correlation of parts and conditions of existence, constituting Cuvier's "natural method" of zoology, offered the politically pious leader of the Paris Museum theoretical motivation for rejecting his colleague Lamarck's theory of species transformation.

Once the transcendental urge was upon a theorist, more abstract unities could be discovered than the four proposed by Cuvier. So in 1820, for instance, Cuvier's colleague Etienne Geoffroy Saint-Hilaire found parallels uniting the mollusca and radiata, on the one hand, and the articulata and vertebrata on the other; he thereby reduced the fundamental types to two.[66] And a decade later, in 1830, he proclaimed a general

65. Georges Cuvier, "Sur un nouveau rapprochement à établir entre les classes qui composent le règne animal," *Annales du Muséum d'Histoire Naturelle* 19 (1812): 73–84. The fourfold classification became the organizing plan for Cuvier's famous *Le règne animal*. See Georges Cuvier, *Le règne animal,* 2d ed., 5 vols., vols. 4 and 5 by P. A. Latreille (Paris: Deterville, 1829–30).

66. Etienne Geoffroy Saint-Hilaire, "Sur une colonne vertèbrale et ses cotes dan les insects apiropodes," *Isis* 2 (1820): 527–52.

FIGURE 9. Baron Georges Cuvier, 1769–1832, fierce opponent of Lamarckian evolutionism; portrait done ca. 1834. Wellcome Institute Library, London.

unity throughout the animal kingdom,[67] something that elevated Cuvier into high academic dudgeon.[68] The assertion that one form of life united the animal kingdom had methodological consequences that Cuvier could not tolerate. It had equally repugnant evolutionary consequences; for if the boundaries between types have been eliminated, rendering all organisms as essentially similar, then gradual transitions among them could be possible. Geoffroy drove through this conceptual opening to advance transformationist proposals echoing those of his colleague Lamarck. Meckel, however, had reached this goal even more quickly.

In the first volume (1821) of his *System der vergleichenden Anatomie,* Meckel argued (as Lamarck had and Darwin later would) that the arbitrary and uncertain distinctions between varieties and species, as well as those between species and the higher classes, suggested the possibility that such "larger and smaller collections of organisms are only alterations, probably originating gradually, of one and the same Urorganism."[69] The structural plan that united the animal kingdom resulted from descent from more primitive stages that commenced with an Urtype. Variations on subsequent types, he thought,[70] might be introduced into species through individuals, namely, by the sort of heritable modifications Lamarck had mentioned. This evolutionary history would be revealed, he argued, in embryogenesis. These parallels between species evolution and individual evolution, moreover, indicated—as they had to his predecessors and would to successors (like Agassiz and Darwin)—that the same developmental laws were at work in both.[71]

---

67. Etienne Geoffroy Saint-Hilaire, *Principes de philosophie zoologique* (Paris: Didier, 1830), preliminary discourse.

68. See Toby Appel's stirring account in her *Cuvier-Geoffroy Debate* (Oxford: Oxford University Press, 1987).

69. Johann Friederich Meckel, *System der vergleichenden Anatomie,* vol. 1 (Halle: Renger, 1821), p. 62.

70. Ibid., pp. 344–45.

71. Ibid., p. 396.

Meckel provided the most sophisticated form of the recapitulation principle, and his work displayed the demonstrative power of comparative embryology for the theory of species evolution. Even Huxley, after Darwin's and Haeckel's arguments for the recapitulation principle finally sank in, came to admit that Meckel's original version, properly understood, conformed to sound biological science.[72] The initial spread of the idea of recapitulation and thus its availability for construing species change, however, was due in no small measure to the embryologist and disciple of Geoffroy, Etienne Reynaud Serres, who regarded Meckel as a venerable "Nestor of medicine."[73] When Serres formulated a version of the recapitulation principle in 1837,[74] Darwin copied it in his notebook and took it to heart.[75] But Darwin's own powerful use of the principle had first to be filtered through the analyses of its most influential antagonist, Karl Ernst von Baer.

Von Baer's Critique of Recapitulation Theory

Von Baer was born into a noble Estonian family and, like many Europeans of his scientific disposition, initially took a degree in medicine, graduating from the small provincial university of Tartu. As a student, he responded to his professors' warnings against *Naturphilosophie* by reading as much as he could; he responded to their lectures with deri-

72. Thomas Henry Huxley, "Evolution," *Encyclopaedia Britannica,* vol. 8, 9th ed. (1878), p. 750: "If Meckel's proposition is so far qualified, that the comparison of adult with embryonic forms is restricted within the limits of one type of organization; and, if it is further recollected, that the resemblance between the permanent lower form and the embryonic stage of a higher form is not special but general, it is in entire accordance with modern embryology; although there is no branch of biology which has grown so largely, and improved its methods so much since Meckel's time, as this."

73. Etienne Reynaud Serres, "Théorie des formations organiques," *Annales des sciences naturelles* 12 (1827): 82–143; quotation from p. 88

74. Etienne Reynaud Serres, "Zoologie: Anatomie des mollusques, *L'Institut, section des Sciences mathématiques, physiques et naturelles* no. 191 (4 January 1837): 370–71.

75. See chap. 5, n. 5.

sion recollected in hostility.[76] Von Baer sought further training at Vienna and there gradually lost his enthusiasm for
medicine. His subsequent study with the exacting Ignaz
Döllinger (1770–1841) at Würzburg ignited a passion for
zoology, particularly comparative anatomy, and rekindled a
smaller flame for Schelling's inspirational vision, which mature years and experience gradually smothered. In 1817 von
Baer took a position at the University of Königsberg, where
Kant had taught. There he began intensive embryological investigations that a decade later yielded to his lasting fame the
discovery of the mammalian ovum. In 1828 he published the
work commonly regarded as initiating modern embryology,
Die Entwickelungsgeschichte der Thiere, which contained his
corrosive critique of recapitulation theory.

Von Baer advanced several objections to recapitulation
theory—that some embryonic stages (e.g., the embryo attached to its yolk) were not met in lower animals; that while
some organs of a lower group might appear similar to those
in the embryo of a higher, the full complement of organs did
not; that some traits of higher organisms appeared as the transitory states in the embryos of lower, and the like. These
objections bore weight, however, only when inserted into
von Baer's theoretical framework, which differed considerably from that of anatomists like Meckel, Treviranus, and
Tiedemann.[77] They assumed continuous progressive transi-

76. Von Baer, *Autobiography,* pp. 80–106.

77. Von Baer's conception, though, exhibits strong resemblances to that of
another recapitulationist, Carl Gustav Carus. Carus remains an elusive historical figure whose ideas seem to have inspired several important zoologists in
addition to von Baer, especially Joseph Henry Green and Richard Owen (see
below for discussions of Green and Owen). Carus, somewhat like Cuvier, argued that the animal kingdom was arranged according to two basic plans,
each with significant subdivisions: the animals without backbone and brain
(the zoophytes, the mollusca, the articulata), and those with backbone and
brain (fish, amphibians, birds, and mammals). An individual organism during embryogenesis would recapitulate lower forms but only within the developmental series corresponding to its particular nature (*Bedeutung*),
which would encompass at least two of the above-mentioned classes. So

tions from lowest organism to highest. Von Baer, however, sided with Cuvier, maintaining that animal life revealed four fundamental arrangements of organic parts, four archetypes (*Haupttypen*) that remained distinct in nature. These archetypes, he granted, displayed different grades of formation (*Ausbildung*) depending on the heterogeneity and differentiation of anatomical structure of species within them. In this respect a bee (a species type within the archetype of the articulata) would represent a higher degree of formation within its archetype than most fish (vertebrates) in theirs. In some comparisons across archetypes and even within them—i.e., down through classes, families, genera, and species—one animal might display one organ system of higher grade of formation than another, but a different organ system of lower grade. The four archetypes represented only four distinct arrangements of organ systems and so could not themselves be placed on a continuum of progressive development. By thus relativizing the idea of progressive scaling, the von Baerean framework frustrated a seemingly fundamental requirement of both recapitulation theory and species evolution.

Individual development, according to von Baer, proceeds from the general features of the archetype, during the early stages of embryogenesis, through higher grades of differ-

---

Carus maintained: "Just as only by progressive stages [*stufenweise*] and in a determinate evolutionary series [*Entwicklungsreihen*] the particular forms of the animal kingdom reach a higher perfection, so must the individual unfold in a certain progressive series. (This proposition has also at times become so misunderstood that some have viewed it as maintaining that, for instance, the mammal fetus must first become a mollusk, then an insect, then a fish, then a bird, etc. They do not consider that in the animal kingdom different evolutionary series are to be found, that it is one progressive series that leads from the zoophyte to the butterfly, another that runs from the same point to slugs, another that goes from the fish to the bird, and another that leads from the fish to the mammal. Thus each individual does not run through all of these developmental series at the same time, but only that appropriate to its being [*Wesen*], to its nature [*Bedeutung*].)" See Carus, *Lehrbuch der Zootomie,* 1st ed., pp. 669–70.

FIGURE 10. Karl Ernst von Baer, 1792–1876; portrait of the great embryologist at age thirty-three

entiation until the particular features of the species are established. The vertebrate fetus does not pass through the permanent forms of the various articulata and mollusca. "The
embryo of the vertebrate is already at the beginning a vertebrate."[78] The human fetus, for example, begins life as a generalized vertebrate and in development passes through the
more specialized type of the (undifferentiated) mammal to
the yet more individualized form of the (undifferentiated)
anthropoid, finally showing specifically human, sexual, and
personal traits. Embryological evolution, in von Baer's view, is
a process of differentiation—a movement from the more homogeneous and universal to the more heterogeneous and individual. It is for this reason that embryos of a given archetype
will resemble, not permanent forms of lower organisms, but
each other at stages prior to attainment of their differentiating
types (i.e., their distinguishing family, genus, or species).

While von Baer rejected the letter of recapitulation theory,
he shared more than enough assumptions with its advocates
that Haeckel, in his *Generelle Morphologie,* could reasonably
call upon von Baer in defense of both species evolution and
individual recapitulation.[79] Like Schelling, whom he studied
while teaching at Königsberg,[80] von Baer understood individual development as a consequence of the essence
(*Wesenheit*) or idea of organization present already in the
just-fertilized egg: "The type of every animal both becomes
fixed in the embryo at the beginning and governs its entire
development."[81] Haeckel, however, could easily translate this

78. Von Baer, *Entwickelungsgeschichte der Thiere,* p. 220.

79. Haeckel, *Generelle Morphologie der Organismen,* 2: 6–12.

80. Von Baer, *Autobiography,* p. 203

81. Von Baer, *Entwickelungsgeschichte der Thiere,* p. 220. Von Baer conceived of this leading principle of embryogenesis—and his minute analysis
upon which it rested—as the ground for confuting the materialist: "Natural
historical research, which one readily assumes promotes and nourishes materialistic views, can itself from observation refute the strict materialistic doctrine and lead to the proof that it is not matter in its exact ordering, but the
essence [*Wesenheit*] (the Ideas according to the new school) that rules the
development of the fetus" (ibid., p. 148). Kenneth Caneva refers to this pas-

power of type governance into the consequences of herita-
bility of previous phyletic forms, and he could interpret the
differentiation of the hierarchy of forms as the result of adaptation in the course of species descent.[82] Von Baer retained
other ideas that also would be turned back to evolutionary
advantage. He recognized, for instance, that the vertebrate
archetype—since it required some organizational features
common to all life—"is, as it were, composed of the earlier
types."[83] This meant that animals of different archetypes
would appear to pass through identical stages, though only in
the earliest periods of their respective developments. It also
meant that creatures of the same archetype but of variously
divergent species would pass through stages so morphologically similar that even von Baer's own trained eye
could hardly distinguish the different embryos:

> The embryo of the mammal, bird, lizard, and snake,
> and probably also the turtle, are in their early stages
> so uncommonly similar to one another that one of-
> ten can distinguish them only according to their size.
> I possess two small embryos in spirits of wine, em-
> bryos whose names I neglected to note down, and I
> am now in no position to determine the classes to
> which they belong. They could be lizards, small
> birds, or very young mammals.[84]

---

sage in his effort to overturn the thesis of Timothy Lenoir, who holds that
"teleomechanists," especially von Baer, thought of the ruling type as but a
function of material organization, not as an independent, nonmaterial force.
See K. L. Caneva, "Teleology with Regrets," *Annals of Science* 47 (1990): 291–
300. The passages from the *Entwickelungsgeschichte* do argue that von Baer
believed the type to be more than material organization. Indeed, von Baer's
entire theory of archetypes resonates of Schelling's *Identitätsphilosophie,*
which would have exemplified for him "the Ideas according to the new
school."

82. Haeckel, *Generelle Morphologie,* 2: 11.

83. Von Baer, *Entwickelungsgeschichte der Thiere,* p. 212.

84. Ibid., p. 221. Darwin would cite this anecdote (based on Huxley's translation of von Baer, see below, chap. 5) but misattribute it to Agassiz, an error
which suggests the recapitulational use he would make of the von Baerean

Perhaps the most significant feature of von Baer's conception that could be assimilated to later evolutionary theory was his admission that within a type the organisms lowest in degree of formation, those least differentiated, would most closely embody the basic type.[85] The archetype could, in a sense, be a real creature and not merely an ideal plan. In this respect, then, the more progressive creatures would indeed pass through some permanent types of those organisms lower in degree of development. These at least were possibilities slumbering in von Baer's considerations. And though he contended against recapitulation theory and the evolution of species, his own conception, having escaped his control, awakened to just these possibilities. This happened first among the English.

---

archetype. See Charles Darwin, *On the Origin of Species* (London: Murray, 1859), p. 439. He corrected his error in later editions of the *Origin,* probably after Huxley pointed it out to him.

85. Von Baer, *Entwickelungsgeschichte der Thiere,* p. 238.

4

EMERGENCE OF
EVOLUTIONARY THEORIES
OF SPECIES CHANGE

From ancient time to the beginning of the nineteenth century, ideas about species change smoldered but failed to ignite the imaginations of most naturalists. Vague notions of transmutation can be traced to classical Greece, where Anaximander (610–546 B.C.), Anaxagoras (534–462 B.C.), and Empedocles (fl. 444 B.C.) concocted fables about the emergence of new sorts of animals that Aristotle (384–22 B.C.) critically analyzed but did not simply dismiss as incompatible with healthy conception.[1] Even the Philosopher acknowledged the creation of new kinds of animals through hybridization; and he elaborated the sort of theory about the spontaneous generation of insects, worms, and marine invertebrates that would later give support to wobbly proposals of species evolution.[2] Indeed, the venerable belief in such "equivocal" productions of ignoble creatures helped convince Francis Bacon (1561–

---

1. Aristotle certainly had objections (*De anima,* 198B. 16–32) to Empedocles' theory that creatures might accidentally have acquired parts and those that proved most viable would survive—a vague prototype of natural selection. However, Aristotle did entertain the possibility that man and the quadrupeds might have originated through spontaneous generation of a grub, which presumably would later develop into the familiar species (*De generatione animalium,* 762B28–763A5). This, as well as much else in Aristotle (e.g., new kinds of animals produced through hybridization—ibid., 746A29–746B12), indicates a need for reevaluating the usual assumption that Aristotelian philosophy could tolerate no consideration of species mutability.

2. See preceding note.

1626) that while "the Transmutation of Species is, in the vulgar Philosophy, pronounced Impossible, . . . seeing there appear some manifest Instances of it, the Opinion of Impossibility is to bee rejected; and the Means thereof to bee found out."[3] Subsequently Benoit de Maillet (1656–1738), Julien Offray de La Mettrie (1709–51), Pierre Louis de Maupertuis (1698–1759), Denis Diderot (1713–84), and Immanuel Kant (1724–1804) all spun gauzy notions about species change.[4] The great natural historian and founder of the Jardin du Roi, Georges Leclerc, Comte de Buffon (1707–88), had initially opposed the idea of transmutation, but breeding experiments and certain theoretical considerations brought him to the view, in his essay "De la dégénération des animaux" (1766), that the originally created kinds of animals (now the genera and families) had, because of the influence of the environment, degenerated into the myriad of species now populating this decidedly ancient earth.[5] Buffon's conviction was not far different from that of his enemy Linnaeus (1707–78), who proposed in his later years the mechanism of hybridization to produce new species from original kinds. Buffon's and Linnaeus's quasi-evolutionary beliefs became known to Erasmus Darwin (1731–1802), who advanced a more radical theory of species alteration in his two-volume Zoonomia (1794–96), according to which that original "living filament" created by God, through irritable response and acquired

3. Francis Bacon, Sylva Sylvarum; Or a Naturell Historie in Ten Centuries, 3d ed. (London: William Lee, 1631), p. 132.

4. For descriptions of transmutational theories before Darwin, see Bentley Glass, ed., Forerunners of Darwin (Baltimore: Johns Hopkins University Press, 1968); and Robert J. Richards, "Influence of Sensationalist Tradition on Early Theories of the Evolution of Behavior," Journal of the History of Ideas 40 (1979): 85–105.

5. Georges Louis Leclerc, Comte de Buffon, "De la dégénération des animaux" (1766), in Oeuvres complètes de Buffon, ed. Pierre Flourens, vol. 4 (Paris: Garnier, 1852–1855). See also Phillip Sloan, "Buffon, German Biology, and the Historical Interpretation of Biological Species," British Journal for the History of Science 12 (1979): 109–53.

habit, became transmuted over eons into all the warm-blooded animals.[6] Charles Darwin (1809–82) first brushed against the hypothesis of species evolution, as he recalled in his *Autobiography*,[7] in reading his grandfather. But undoubtedly the equally important early influence on Darwin, as well as the singular stimulus for his contemporaries (especially Robert Grant, Herbert Spencer, Robert Chambers, and Ernst Haeckel), was Jean-Baptiste de Lamarck (1744–1829).

Darwin had learned of Lamarck's hypothesis under the tutelage of Robert Grant (1793–1874), who directed him in a vocational study of invertebrates while he was a very young medical student at Edinburgh (1825–27); and during the voyage on board HMS *Beagle* (1831–36), Darwin had leisure to examine Lamarck's *Histoire naturelle des animaux sans vertèbres* (1815–22). In 1832, when sailing down the eastern coast of South America, he received the second volume (1832) of Charles Lyell's *Principles of Geology* (1830–33). That book contained a sustained presentation and sharply negative critique of Lamarck's arguments "in favour of the fancied evolution of one species out of another."[8]

Lamarck had first suggested species change in 1800 but developed his theory more fully in later works, especially in his *Philosophie zoologique* (1809). It was this treatise that Lyell analyzed. Lamarck proposed that simple monadic life bubbled up from the muck under the influence of the imponderable fluids of caloric and electricity. These forces, which continued to operate even today, caused those elementary vesicles born of slime to become more complex, eventually producing organisms that internalized the fluids. The incessant excavations and articulations by the internal fluids over

6. Erasmus Darwin, *Zoonomia or the Laws of Organic Life,* 2d ed. (London: Johnson, 1796), 1: 509.

7. Charles Darwin, *The Autobiography of Charles Darwin,* ed. Nora Barlow (New York: Norton, 1969), p. 49.

8. Charles Lyell, *Principles of Geology* (London: Murray, 1830–33), 2: 60.

Ambroise Tardieu direxit.

FIGURE 11. Jean-Baptiste de Lamarck, 1774–1829; engraving in 1821. Wellcome Institute Library, London.

generations led to greater perfection (i.e., complexity) of organisms, while the heritable effects of habit modified their parts to fit them into a changing environment.[9] Unlike Treviranus, Tiedemann, and later Darwin, Lamarck did not appeal to evidence of great extinctions of animals to argue for transformations in organic life. The fossil evidence that Cuvier accumulated had already been piously wedded to the geological theory of catastrophism, while Lamarck cast his lot with uniformitarianism—the theory that geological formations resulted from processes occurring slowly over vast reaches of time instead of being wrenched into place through forces ultimately controlled by an unpredictable Deity. Seemingly extinct creatures, he urged, had been transformed into contemporary species, or might even be found in unexplored regions of the globe.[10]

Lamarck did not use the term "evolution" to describe the progressive transformation of animal species; rather he referred to the process, in obvious deference to his mentor Buffon, in a negative way, as the inverse of that "degradation [*dégradation*]" we observe as we cast our eye down the scale of life.[11] Lyell gave currency to the English usage of the term "evolution" by applying it indifferently to Lamarck's theory of transmutation and Tiedemann's idea that during embryogenesis the brain of higher animals evolved through stages of the lower. Lyell, however, seems to have gotten the sterling for his

9. For discussions of Lamarck's theory, see Richard Burkhardt, *The Spirit of the System* (Cambridge: Harvard University Press, 1977); Pietro Corsi, *The Age of Lamarck* (Berkeley: University of California Press, 1988); and Robert J. Richards, *Darwin and the Emergence of Evolutionary Theories of Mind and Behavior* (Chicago: University of Chicago Press, 1987).

10. Jean-Baptiste de Lamarck, *Systéme des animaux sans vertébres* (Paris: Lamarck et Deterville, 1801), p. 407. Lamarck thought the hypothesis of general geological upheavals "a rather convenient means for those naturalists who wish to explain everything and who do not take any trouble to observe and study the course which nature follows in regard to her productions and all that constitute her domain" (ibid.).

11. Jean-Baptiste de Lamarck, *Philosophie zoologique* (Paris: Dentu, 1809), 1: 220.

FIGURE 12. Sir Charles Lyell, 1797–1875, who wrote an extended critique of Lamarck's and Tiedemann's evolutionism; lithograph of portrait done in 1849. Wellcome Institute Library, London.

coinage from Serres, as well as perhaps from Joseph Henry Green and fellow Scot Robert Grant.[12]

Serres, Grant, Green, and Lyell on Recapitulation and Evolution

I have argued that the principle of recapitulation transmitted to the idea of species alteration certain considerations born out of embryological theory. The term "evolution" indexes this transmission. The Germans, especially Tiedemann and Meckel, thought of the principle of recapitulation as reflecting a unity of law that accounted for both individual development and species development. In his own work on recapitulation, which owed much to the Germans, Serres used the expression *théorie des évolutions*[13] ambiguously to refer to the recapitulational *métamorphoses* of organic parts in the individual and the parallel changes one sees in moving (intellectually) from one family of animals to another and from one class to another. A similar usage was introduced to English audiences at about the same time by two medical zoologists of very different theological and metaphysical persuasions—Robert Grant, in 1826, and Joseph Henry Green, in his Hunterian Lectures for 1827 and 1828.

While at medical school in Edinburgh Darwin made the acquaintance of Robert Grant (1793–1874), who directed him in the investigation of marine invertebrates. In his *Autobiography* Darwin recalled that Grant "one day, when we were walking together burst forth in high admiration of Lamarck and his views on evolution." Darwin reacted, he remembered, "in silent astonishment"; but added, perhaps in

12. Adrian Desmond believes Robert Grant to have been the first to introduce the term "evolution" in its species-transforming sense into English print. See his admirable *Politics of Evolution* (Chicago: University of Chicago Press, 1989), p. 5. Desmond's ascription to Grant of the anonymous journal article of 1826, in which the word appeared, is not unproblematic. See n. 17, below.

13. Etienne Reynaud Serres, "Théorie des formations organiques," *Annales des sciences naturelles* 12 (1827): 83.

defense of his own accomplishment, that the encounter was
"without any effect on my mind."[14] Grant, who received his
M.D. from Edinburgh in 1814, traveled through Germany and
lingered in France, where, despite Cuvier's towering authority, he became an admirer of Lamarck and a friend of
Geoffroy. In 1827 Grant was appointed to the chair of comparative anatomy at the new University of London after the
city council refused to meet the monetary demands of the
very German Professor Johann Friedrich Meckel, to whom
the position had first been offered. In his university capacity,
Grant delivered several series of lectures in which recapitulation became the central theoretical focus.[15] These lectures
might only have hinted at transformationism, but even that
would have been obscured by his exceedingly dour style,
which was characterized for Darwin by a friend as "rather too
grave, & rather too pedantic, too much given to coin hard
words, at times . . . eloquent & animated, generally verbose
& lengthy."[16] He was more forthright as a younger man. In an
early article (1826), published under a discreet veil of ano-
nymity, Grant declared for Lamarck. The French zoologist, according to Grant, had found "a key to the profoundest secrets
of nature," namely, that infusoria and worms had spontaneously been generated and "that all other animals, by the

14. Darwin, *Autobiography,* p. 49.

15. Beginning in 1833, Grant delivered a series of some sixty lectures at
University College on the theme of animal unity. As he considered each organ
system of man, he would conclude by pointing to comparable organs in
lower animals, which were, he maintained, recapitulated in human embryogenesis. He summed up his position in the final lecture on mammals:
"The transient forms of the organs in the higher classes of animals represent
successively the permanent forms of the lower, and the laws of animal development have continued to operate with the same undeviating uniformity
since the first appearance of animal forms on the earth, as we observe in
those which govern the rest of the material world." See the report of this lecture "On the Generative System of Mammalia," *Lancet* 2 (1833–34): 1035.

16. F. W. Hope to Charles Darwin (15 January 1834), *Correspondence of
Charles Darwin,* ed. Frederick Burkhardt et al. (Cambridge: Cambridge University Press, 1985–), 1: 363.

FIGURE 13. Robert Edmond Grant, 1793–1874, whose praise for Lamarck surprised his young friend Darwin; lithograph in 1852. Wellcome Institute Library, London.

operation of external circumstances, are evolved from these in a double series, and in a gradual manner."[17] Though Lamarck had not engaged recapitulation theory as part of his conception, it yet became for Grant—probably because of the influence of Geoffroy and Serres—another magical key. It unlocked the secret unity that spread throughout the animal kingdom and revealed the common set of laws that made such unity comprehensible, without, he would argue to friends and insinuate in public lectures, the need of appeal to a meddlesome Deity. Desmond, who has provided the most thorough and subtle account of Grant's life, contends that this Scots materialist vividly illustrated for the more orthodoxly sober, like Richard Owen, the cautionary implications of evolutionary theory.[18] Despite, however, the radical dangers to religion and social polity that evolutionary theory might have suggested to some, it could be tamed, as Joseph Henry Green (1791–1863) demonstrated.

Green, who became professor of surgery at the new King's College in 1830, had been educated in Germany and cultivated a knowledge of transcendental anatomy, especially the works of Oken, Tiedemann, and Carus.[19] Green was encouraged in this philosophical-physiological pursuit by his friend-

17. [Robert Grant], "Observations on the Nature and Importance of Geology," *Edinburgh New Philosophical Journal* 1 (1826): 293–302; quotation from p. 297. Given the subject of the article, the editor of the journal (the geologist Robert Jameson), and the source of the publication, Lyell very likely read Grant's piece and would have seen the word "evolved" used in description of Lamarck's theory. The established view that Grant authored this essay is criticized by James Secord, who argues with dexterous contextuality that the piece was more likely written by the editor Jameson. See James Secord, "Edinburgh Lamarckians: Robert Jameson and Robert E. Grant," *Journal of the History of Biology* 24 (1991): 1–18. While no fast facts fix the identity of the author, the essay certainly expresses a series of views close to ones Grant held.

18. See Desmond, *Politics of Evolution,* pp. 276–79.

19. Phillip Sloan discusses Green's background in the masterful "Introduction" to his edition of *Richard Owen's Hunterian Lectures, May–June 1837* (London: British Museum [Natural History]; Chicago: University of Chicago Press, forthcoming). I am grateful to Sloan for calling Green's work to my attention. See also Desmond, *Politics of Evolution,* pp. 260–75.

ship with Samuel Taylor Coleridge (1772–1834), who knew
thoroughly the views of Erasmus Darwin, Treviranus, and
Tiedemann. Indeed, Green's Hunterian lectures of 1827
and 1828 were tinctured with Coleridgean "evolutionary"
individualism—an elixir later stimulating Herbert Spencer
(1820–1903) as well.[20] In these lectures, Green suggested to
his listeners (among whom was his protégé Richard Owen,
and perhaps Lyell)[21] that nature, in producing the great types
of animal organization, labored "to complete in the evolution

20. Spencer is usually credited with giving currency to the word "evolution" in its use to describe species change. As we have seen, and more evidence will be provided below, the term was commonly used with this
meaning prior to Spencer's influential marketing of it. Spencer seems to have
first employed the word in 1851, in *Social Statics,* where he argued that the
moral law, or the law of equal freedom, is the same principle that governs the
organic development of men toward ever greater individuality—"an endowment," he claimed, "now in process of evolution." See Herbert Spencer,
Social Statics, or the Conditions Essential to Human Happiness (London:
Chapman, 1851), p. 440. Spencer borrowed, as he acknowledged, this conception of life as a progressive evolution toward individuality from Samuel
Taylor Coleridge's posthumous essay (1848) "The Theory of Life." See
Samuel Taylor Coleridge, *Miscellanies, Aesthetic and Literary: To Which is
Added the Theory of Life,* ed. T. Ashe (London: Bell & Sons, 1892). Coleridge
used "evolution" to describe the tendency of life to attain progressively
greater individuality—a view he in turn seems to have adopted from Schelling, whose works he knew extensively and intimately. Coleridge's evolution,
like Schelling's, bespoke only ideal relations of species, not historical relations, a metaphysical nicety that probably escaped the young Spencer.
Spencer would also have met the term "evolution" in its "species-
transformation" sense in Lyell's *Principles of Geology* and Carpenter's *Principles of Comparative Physiology* (see below), both of which he read prior to
composing *Social Statics.* See Richards, *Darwin and the Emergence of Evolutionary Theories,* pp. 267–69, for a discussion of Spencer's use of Lyell and
Carpenter.

21. Lyell had undertaken serious study of Lamarck's *Philosophie zoologique* in late summer of 1827, while in Scotland. He returned to London the
following January and remained there until his trip to Paris in May. For these
months, see Leonard Wilson, *Charles Lyell: The Years to 1841* (New Haven:
Yale University Press, 1972), pp. 184–90. Green, who had considerable admiration for Lamarck despite certain reservations about the Frenchman's theory, began his lectures that March. Thus Lyell could have attended and would
have been motivated to do so. See n. 23 below.

of the organic realm" her highest Idea, namely, man. This same progressive tendency expressed itself in the parallel evolution of the embryo: "Hence the states, which the individual passes through in all the epochs of its embryonic being, and which having been disappear, are preserved in Nature, and maintain the rank of external and abiding forms."[22] Green hastened to add that this "series of evolutions from the lowest to the highest" occurs not linearly, but as a tree whose "infinitely diversified [branches] manifest the energy of the life within." Nor did the productive force operate according to "the fanciful scheme" of Lamarck;[23] rather, "the ascent is

22. Joseph Henry Green, "Recapitulatory Lecture" (1828), in his *Vital Dynamics: The Hunterian Oration before the Royal College of Surgeons in London, 14th February 1840* (London: Pickering, 1840), p. 103. This line comes directly from Tiedemann (see text to n. 54, chap 3, above). The "Recapitulatory Lecture" was a transcription of Green's Hunterian Lecture for 1828. Owen was present for the 1828 lecture; see Sloan's introduction to *Richard Owen's Hunterian Lectures*. Green's Hunterian Lecture for the previous year, 1827, expressed these same "evolutionary" ideas in virtually the same terms. See n. 24 of this chapter for the representative quotation. In his Hunterian Lecture of 1840, Green credited Hunter with first formulating the principle of recapitulation—see n. 3, chap. 3, above. Green met the principle, however, also in the work of many of the German authors he knew intimately, particularly Tiedemann and Carus. In his *Lehrbook der Zootomie,* which Green used as a textbook, Carus not only advanced a cautious recapitulation theory (see nn. 8 and 77, chap. 3, above) but also maintained that the "evolution of nature [*Natur in ihrer Entwicklung*]" indicates a unity in the animal kingdom, namely, that constituted by "the Idea of Animality, which has its highest expression in the human organism." See Carl Gustav Carus, *Lehrbuch der Zootomie* (Leipzig: Gerhard Fleischer the Younger, 1818), p. 6.

23. I have no direct evidence that Lyell was familiar with Green's work; but just like Green, Lyell also characterized Lamarck's ideas as a "fancied evolution" (see text to n. 8 in this chapter), which is at least mildly indicative of an intellectual connection. Further, Green, in his "Recapitulatory Lecture" (p. 122), expressly considered, as did Lyell, Tiedemann's evolutionary theory: "If the facts in question were evidence less decisive of a process of development, the deficiency would be abundantly supplied by the curious researches of Tiedemann, on the formation of the foetal brain (*Bildungsgeschichte des Gehirns*). In tracing the evolution of the brain, he has satisfactorily shown the correspondence of the temporary stages of its construction in the *foetus* to the permanent forms of the organ characterizing the inferior classes."

the indication of a law, and the manifestation of a higher power acting in and by nature."[24]

Desmond argues, almost convincingly, that Green rejected the Lamarckian notion of a lower species being transformed into a higher. He thinks that Green denied species evolution principally because of his Coleridgean class politics. That political attitude, according to Desmond, decreed a stately hierarchy of social orders, while in Green's nostrils the descent thesis reeked of political radicalism. Green must have believed, Desmond reckons, that the spoiled scientific carrion of evolution could only incite the democratic levelers.[25]

The student of history suspicious of a priori schemes, such as those advanced by social constructionists, will forbear immediate acceptance of Desmond's thesis for at least two rea-

24. Green, "Recapitulatory Lecture," pp. 108–9. This lecture repeated in almost identical terms the evolutionary perspective he developed in his 1827 lectures on the "Natural History of Birds." In that earlier presentation he reminded his audience that "we have . . . set out upon the plan of considering nature as a series of evolutions, from the lower, from the lowest form in which Life manifests its power in a production of animated being up to its most complex forms—& I presented to you this view not under the idea that the lower had any power of assuming the rank and privileges of the higher nor upon any such fanciful scheme as that which that otherwise most meritorious naturalist Lamarck has proposed for the invertebrated series of animals, but as the lower passing by a series of evolutions to the higher, under the law & influence of a higher power acting in and by nature." See Sloan's transcription of Green's "Introductory Lecture to the Natural History of the Birds, 1827," included as an appendix to his *Richard Owen's Hunterian Lectures*.

25. Desmond makes a wonderfully contextualized case for deeply held political motives determining the scientific debate over evolutionary theory prior to Darwin's *Origin of Species.* My principal complaint, which touches the heart of his thesis, concerns his insistence that political motives always crawl just beneath the covers of a scientific proposition. While scientific theories—or historiographic ones for that matter—may be moved principally by the lower life of political partisanship, they may have other stimulating causes, for instance, reason and evidence, which may be embedded in scientific tradition, philosophical conviction, or religious belief. The movement under the spread need not always be politics; sometimes scientific desires resonate to their own particular loves. For Desmond's analysis of Green's motivation, see *The Politics of Evolution,* pp. 260–75.

FIGURE 14. Joseph Henry Green, 1791–1863, an early English evolutionist and transcendental morphologist, and mentor of Richard Owen; lithograph of him at age fifty-five. Wellcome Institute Library, London.

sons. First, he claims that Green, in rejecting "the fanciful scheme of Lamarck," also denied a historical transformation of species. But Green's lectures of 1827 and 1828 indicate clearly that he demurred, not to the idea of genealogical descent, but to two attendant features of Lamarck's scheme, namely, the proposals for the mechanism and for the mode of descent. Green denied Lamarck's "inherent" natural source of species change, since the Frenchman's theory was that "in which the ground and cause is everywhere meaner and feebler than the effect." Green rather insisted that "the ascent is the indication of a law, and the manifestation of a higher power acting in and by nature."[26] He also maintained against Lamarck that "this gradation and evolution of animated nature is not simple and uniform; [rather] nature is ever rich, fertile, and varied in act and product; [like] some monarch of the forest."[27] Evolution, he argued, spread through great tree-like ramifications. Thus it was not species transmutation per se to which Green objected but to the materialistic and simplified model of that process. Whether he actually believed that one species gave birth to another, even under divine edict, would be difficult to say, since his transcendentalist language slides over those clear distinctions that post-Darwinians require. Certainly there is nothing in his Hunterian lectures that excludes the idea of a gradual transmutation; indeed, given his familiarity with Lamarck's and Tiedemann's ideas and his use of the language of "ascent . . . by nature," a straightforward interpretation would have him endorsing progressive species alteration. Like Grant and the Germans whom they both read, Green thought of progressive ascent in embryo and species as comparable transformations that expressed a unity of law governing the living world—and for Green such biological symmetry had the added purpose of lifting our spirits to the Divine. Green's shaded remarks on transmutation but clear perspective on

26. Green, "Recapitulatory Lecture," p. 108.
27. Ibid., p. 109.

the originating sacred cause of species would later be reflected in the ambiguous but safe position of his protégé
Richard Owen. Yet just as in the case of Owen (see below), it is
obvious that Green did not quite formulate his lectures with
our precise question in mind. Rather, as a Germanophilic
teleologist of religious passion, he desired to evacuate the
possibility that matter alone could rouse itself to more perfect productions.

But even if we drop the question of whether Green was
really a species evolutionist—he was certainly an embryological evolutionist who endorsed "Tiedemann,[28] who
was a species evolutionist—a second consideration needs to
be introduced, which leads to the vital center of Desmond's
thesis. This concerns the priority of motivation that he has assumed. Desmond has set the friends of Coleridge against the
Lamarckians because of class politics rather than because of
the intellectual commitment to a certain kind of metaphysics
of nature. But one must remember that the Coleridgean belief in the ascent of the individual fed the Lamarckian evolutionism of that anarchistic class smasher Herbert Spencer,[29]
and at the same time it succored the enlightened radicalism
of John Stuart Mill.[30] Most scientific theories are politically in-
sipid. The same theory can be consumed by thinkers of very
different social ideals (e.g., the evolutionism taken in by
Spencer and Marx, or Wilson and Gould). So a particular theory may be chosen for reasons other than a distinctive political flavor. Green perhaps inclined toward the aristocratic, but
his Platonic and teleological impulses, along with his detailed
knowledge of comparative anatomy—of the sort which led
his German guides to species transformation—would have
been enough to justify a theory that had nature striving toward greater individuality and consciousness, culminating in

28. See nn. 22 and 23, above.

29. See Richards, *Darwin and the Emergence of Evolutionary Theories of
Mind and Behavior,* pp. 247–60; and also n. 20, above.

30. See Mill's essay on Coleridge, in *John Stuart Mill, On Bentham and
Coleridge,* introduction by F. R. Leavis (New York: Harper Torch, 1962).

the precast ideal of rational humanity. A perfectly condign analysis might have Green adopting his theory of the law-governed development of species independently of any political conviction, the reason being its congenial metaphysical implications and the empirical evidence supporting it. Such an analysis, though, might subsequently show how Green then used his science to support a favored political position. Any assumption of the priority of the political motive in determining a scientific stand may vent a late twentieth-century political-historical passion more than an early-nineteenth-century natural-historical one.

But enough (perhaps more than enough). Green's importance—for this historian at least—lies in his use of the term "evolution" to describe, and thus implicitly to identify, the development of the embryo and the parallel development of species and in his formulation of the Germanic notion that a common law, for him a divine command, governed this dual evolution.

Lyell never mentioned Grant or Green in the *Principles of Geology,* though the likelihood of his knowing their works was great.[31] He did, however, cite Serres and refer to Tiedemann in evaluating the evidence which the principle of recapitulation provided for a real evolution of species.[32] Lyell thought the principle really only demonstrated a Cuvierian or Geoffroy-like unity of plan within the animal kingdom—transcendental relations but no real historical links. Nonetheless, by use of the term "evolution" he suggested to a wide readership the intellectual connection between the theories

31. See nn. 17, 21, and 23, above. While Lyell was composing the second volume of his *Principles,* which contained his analysis of Lamarck, he was simultaneously pursuing a professorship at Kings College, London (1831), where Green was the Hunterian professor. See Wilson, *Charles Lyell,* pp. 308–28, for a discussion of Lyell's efforts to secure a professorship.

32. Lyell, *Principles of Geology,* 2: 62–64. Lyell uses the term "evolution" to refer to both Lamarck's theory of species transmutation (p. 60) and Tiedemann's thesis that during gestation the brain of higher creatures "evolved" through stages of the lower (p. 63).

of individual (that is, embryological) evolution and species evolution. One of those who very carefully dissected Lyell's remarks in the *Principles of Geology* was Darwin.

Darwin's Theories of Species Change

After returning from the *Beagle* voyage in October 1836, Darwin began arranging and cataloging his specimens. In March of the next year he set down in his "Red Notebook" brief speculation on species change.[33] He further reflected on such changes in a series of notebooks beginning in late spring or early summer 1837, the "Transmutation Notebooks." In September 1838 he read Malthus, which, as he recalled, gave him "a theory by which to work."[34] The catalyst of Malthus precipitated out of Darwin's thought the bare structure of his mechanism of natural selection. He further developed his ideas about species change in two essays, one in 1842 and its expansion in 1844. These became the spine for the more comprehensive expression of his theory, which he began drafting in 1856. The work on this book, which was to be called "Natural Selection," was interrupted in 1858 by a letter from Alfred Russel Wallace (1823–1913) that outlined a theory of transmutation very similar to the one Darwin had been laboring over for twenty years. In a white heat, he condensed his huge manuscript and added in smaller compass the further chapters he intended. He published the *Origin of Species* in 1859. Through this long gestation, one can distinguish certain stages in the development of his theory.[35]

33. For historical analyses of Darwin's first formulations of ideas of species change, see Sandra Herbert's introduction to *The Red Notebook of Charles Darwin,* ed. Sandra Herbert (Ithaca: Cornell University Press, 1980), pp. 11–12; Frank Sulloway, "Darwin's Conversion: The *Beagle* Voyage and Its Aftermath," *Journal of the History of Biology* 15 (1982): 325–96; and M. J. S. Hodge, "Darwin and the Laws of the Animate Part of the Terrestrial System (1835–1837)," *Studies in the History of Biology* 6 (1983): 1–106.

34. Darwin, *Autobiography,* p. 120.

35. For analyses of the stages of Darwin's intellectual development, see David Kohn, "Theories to Work by: Rejected Theories, Reproduction, and

In the earliest stage,[36] Darwin—reflecting ideas of his grandfather and Lyell—considered species to be comparable to individuals: under the influence of the environment, species changed over time; but reaching the end of their allotted years, they gave birth to new species and then died off. In the second stage, Darwin retained the adapting mechanism of the heritable effects of environmental agents (taking many hints from his grandfather and Lamarck) but gave up the idea of species having a definite term of life. He supposed that there would be a branching of species, with some continuing to "progress" and "perfecting" through different forms, while others, because of not adapting fast enough, would go extinct.[37] This, Darwin thought, was comparable to what occurred when individuals, having adapted to their circumstances, produced offspring with the new modifications; the parents, however, had to die so that progeny could take their place: "generation of species like generation of individuals.—Why does individual die, to perpetuate certain peculiarities, (therefore adaptation) . . . Now this argument applies to species.—If individual cannot procreate, he has no issue, so with species."[38] Like his grandfather, Grant, Green, and the German transcendentalists, Darwin initially under-

Darwin's Path to Natural Selection," *Studies in the History of Biology* 4 (1980): 67–170; M. J. S. Hodge, "Darwin and the Laws of the Animate Part of the Terrestrial System," and "Darwin as a Lifelong Generation Theorist," *The Darwinian Heritage,* ed. David Kohn (Princeton: Princeton University Press, 1985); and Richards, *Darwin and the Emergence of Evolutionary Theories of Mind and Behavior,* pp. 83–105.

36. Darwin's initial reflections on species change are recorded in his "Red Notebook," MS pp. 127–30, and "Notebook B," MS pp. 1–23. These and his other early notebooks have been meticulously transcribed and edited in *Charles Darwin's Notebooks, 1836–44,* ed. Paul Barrett, Peter Gautrey, Sandra Herbert, David Kohn, and Sydney Smith (Ithaca: Cornell University Press, 1987). The cited references come from pp. 61–62 and 170–76 of the Barrett et al. edition.

37. Darwin, "Notebook B," MS pp. 25–39 (Barrett et al., pp. 177–81).

38. Darwin, "Notebook B," MS pp. 63–64 (Barrett et al., p. 187).

stood the goal of species transformation to be higher creatures, ultimately man: "Progressive development gives final
cause for enormous periods anterior to Man."[39]

In the third stage of his reflections, Darwin realized that the
mechanism of the direct effects of the environment would
not easily adjust organisms to their surroundings. He then
proposed that an animal would develop new habits, which
would adapt it to a shifting environment, and that these habits
would over time be sifted, so that individual peculiarities
would fall out. Finally, by dint of repetition over countless
generations, such general habits would become instinctive.
These finely articulated innate behaviors would in their
turn slowly alter anatomical traits, changing the character of
species:

> an action becomes habitual is probably first stage, &
> an habitual action implies want of consciousness &
> will & therefore may be called instinctive.—But why
> do some actions become hereditary & instinctive &
> not others.—We even see they must be done often
> ⟨⟨to be habitual⟩⟩ or of great importance to cause
> long memory.—structure is only gained slowly.—
> therefore it can only be those actions, which Many
> successive generations are impelled to do in same
> way—The improvement of reason implies diversity
> & therefore would banish individual, but general
> ones might yet be transmitted.[40]

By postulating gradual modifications introduced by instinct,
Darwin constructed his mechanism to Lyellian uniformi-
tarian specifications, while rejecting what he took to be the
Lamarckian attribution of conscious will effort to animals.[41]

In late September 1838, Darwin read Malthus's *Essay on*

39. Darwin, "Notebook B," MS p. 49 (Barrett et al., p. 182).

40. Darwin, "Notebook C," MS p. 171 (Barrett et al., p. 292). Single wedge
quotes indicate Darwin's deletions; double wedge quotes indicate his
insertions—the standard convention that will be used from here on out.

41. See Richards, *Darwin and the Emergence of Evolutionary Theories of
Mind and Behavior,* pp. 85–98.

Population, which brought him to appreciate the sort of pressure to adapt that would be produced by great fecundity. In this light he immediately understood the advantage a favorable trait would have and how it might gradually alter species:

> [Sept.] 28th. ⟨⟨I do not doubt, every one till he thinks deeply has assumed that increase of animals exactly proportional to the number that can live.—⟩⟩ We ought to be far from wondering of changes in number of species, from small changes in nature of locality. Even the energetic language of ⟨Malthus⟩ ⟨⟨Decandoelle⟩⟩ does not convey the warring of the species as inference from Malthus . . . [I]n Nature production does not increase, whilst no checks prevail, but the positive check of famine & consequently death . . . One may say there is a force like a hundred thousand wedges trying force ⟨into⟩ every kind of adapted structure into the gaps ⟨of⟩ in the oeconomy of Nature, or rather forming gaps by thrusting out weaker ones. ⟨⟨The final cause of all this wedgings, must be to sort out proper structure & adapt it to change.⟩⟩[42]

Malthus brought Darwin to see that under increasing population pressure, the features of local environments would select, as it were, those individuals that happened to have traits which could wedge them into unoccupied parts of their surroundings—or which would give them an advantage over competitors already inhabiting particular spaces. Those with the favorable traits would live to propagate another day, while those bested would untimely perish, leaving few behind them. The selected organisms would thus be able to transmit their advantageous traits to offspring and so deliver their competitive virtues down the generations. The resulting proportional increase of those bearing the superior traits would, in the measure of time, gradually transform the species.

In the reflections of 28 September, Darwin's mechanism of

42. Darwin, "Notebook D," MS pp. 134e–35e (Barrett et al., pp. 374–75). See n. 40.

natural selection came to birth. When it did, it slowly pushed the other devices of species change to the periphery of his theory of evolution. Though he regarded these "Lamarckian" instruments as no longer central to the production of new species, he did retain them as various auxiliary mechanisms in the *Origin of Species*. So, for instance, anatomical changes produced through acquired habit could, as he yet maintained in the *Origin,* become impressed upon the hereditary substance and be passed to succeeding generations; as well, such acquired characteristics might also serve as variations upon which natural selection could operate.[43] Despite the preservation of these earlier devices of species change, Darwin urged in the *Origin* that natural selection accomplished most of the work done in evolution.

Natural Selection as the Mechanism of Progressive Evolution

Natural selection appears to modern eyes to foredoom any notions of evolutionary progress. After all, Darwin's mechanism works on chance variations, which it selects to satisfy local requirements. Indeed, Gould, in his interpretation of Darwin, has pounded natural selection into a fine, anti-progressivist blade, which he has wielded to protect the lineage of neo-Darwinism. Gould maintains that "an explicit denial of innate progression is the most characteristic feature

43. In the *Origin,* Darwin listed use and disuse among the sources of variation for selection. He said: "From the facts alluded to in the first chapter, I think there can be little doubt that use in our domestic animals strengthens and enlarges certain parts, and disuse diminishes them; and that such modifications are inherited." See Charles Darwin, *On the Origin of Species* (London: Murray, 1859), p. 134. He also asserted that acquired habit, like selection, could act more immediately to introduce adaptations; but he yet thought his primary device would always retain the upper hand (ibid., pp. 142–43): "On the whole, I think we may conclude that habit, use, and disuse, have, in some cases, played a considerable part in the modification of the constitution, and of the structure of various organs; but that the effects of use and disuse have often been largely combined with, and sometimes overmastered by, the natural selection of innate differences."

separating Darwin's theory of natural selection from other nineteenth century evolutionary theories. Natural selection speaks only of adaptation to local environments, not of directed trends or inherent improvement.[44] Certainly historians of this mind have the avowal of Darwin himself, when he protested against Lamarck's idea of an "innate tendency toward progressive development."[45]

Scientifically sustained ideas of progress appear, from our late-twentieth-century perspective, to be tainted by foreign political agents. Within the Victorian environment of such ideas, natural selection *as we now understand it* might appear to be a deduction decidedly dangerous. Gould explains:

> Its [natural selection's] Victorian unpopularity, in my view lay primarily in its denial of general progress as inherent in the workings of evolution. Natural selection is a theory of local adaptation to changing environments. It proposes no perfecting principles, no guarantee of general improvement; in short, no reason for general approbation in a political climate favoring innate progress in nature.[46]

The historian, and especially the historian-scientist, can, I believe, become too easily beguiled by the power of present scientific theory and consequently imagine that its ancestor theory carried the same "logical" implications, which are then presumed to have stood clear to the earlier practitioners. It is convenient to forget that scientific theories, like

44. Stephen Jay Gould, "Eternal Metaphors of Palaeontology," *Patterns of Evolution as Illustrated in the Fossil Record,* ed. A. Hallan (New York: Elsevier, 1977), p. 13.

45. Charles Darwin to Alpheus Hyatt (4 December 1872), *More Letters of Charles Darwin,* ed. Francis Darwin (London: Murray, 1903), 1: p. 344; see similar remarks in Darwin's "Essay of 1842," transcribed in *The Foundations of the Origin of Species: Two Essays Written in 1842 and 1844 by Charles Darwin,* ed. Francis Darwin (Cambridge: Cambridge University press, 1909), p. 47.

46. Stephen Jay Gould, *Ever Since Darwin* (New York: Norton, 1977), p. 45.

biological species, are also historical entities whose logic must be contingently read. While Darwin did reject the hypothesis of an *intrinsic cause of necessary progress* buried in the interstices of organization, in the beginning he nonetheless insisted, relying on the embryological model, that animals had an internal "tendency to change,"[47] which would be progressively molded by the *extrinsic* agency of the environment. Natural selection would exert, as it were, an external pull, drawing most organisms to greater levels of complexity and perfection.[48]

The magnetic attraction to perfection would be, Darwin thought, induced by the peculiar dynamic of selection. He supposed that the environment against which organisms were most often selected would be the living environment of other creatures, so that reciprocal developmental responses would be evoked throughout the system. It might be, of course, that extinction of simpler animals in a particular location would lead to some more advanced creatures becoming simplified, to backfill the gap. This would mean that some animal series might show a kind of devolution, but the trend would nonetheless be toward ever-increasing complexity. Darwin expressed his belief in a progressive, natural selection dynamic in his fourth transmutation notebook, some time in January of 1839:

> The enormous number of animals in the world depends on their varied structure & complexity.— hence as the forms became complicated, they opened *fresh* means of adding to their complexity.—but yet there is no *necessary* tendency in the simple animals to become complicated although all perhaps will

47. Darwin, "Notebook B," MS pp. 5, 16, 18, 20 (Barrett et al., pp. 171, 175).

48. I have analyzed Darwin's use of natural selection as a vehicle of biological progress in Robert J. Richards, "The Moral Foundations of the Idea of Evolutionary Progress: Darwin, Spencer, and the Neo-Darwinians," *Evolutionary Progress,* ed. Matthew Nitecki (Chicago: University of Chicago Press, 1988). See Hodge, "Darwin and the Laws of the Animate Part of the Terrestrial System," for a complementary analysis.

have done so from the new relations caused by the advancing complexity of others.—It may be said, why should there not be at any time as many species tending to dis-development (some probably always have done so, as the simplest fish), my answer is because, if we begin with the simpler forms & suppose them to have changed, their very changes tend to give rise to others.[49]

Darwin's notion of an innate tendency to change gradually faded in his theory; it came to be replaced by the supposition of environmental forces producing the kind of variation that could be transmuted into progressive forms during the development of species. (Darwin did argue in the *Origin* that organisms, especially those in large genera, would inherit a tendency to vary and diverge from parent forms—the residual of his idea of an innate tendency to change.)[50] He retained, however, his conception of a progressive dynamic; and in the *Origin of Species* he integrated that idea into the more general theory of divergent evolution. In large, open areas the environment of other closely related species would promote mutually adaptive lineages, the species of which would, as a result, continuously diverge and improve; for in such situations "the conditions of life are infinitely complex from the large number of already existing species; and if some of these many species become modified and improved,

49. Darwin, "Notebook E," MS p. 95 (Barrett et al., pp. 422–23).

50. Darwin, *Origin of Species,* p. 118. Darwin made the idea of a heritable tendency to vary a part of his more general theory of divergence. He believed that the theory of divergence, which he slowly formulated during the years 1854–57, was crucial for explaining the branching character of evolutionary descent. After all, evolution could have proceeded linearly. In a biologically shifting environment, in which new niches were always opening up, individuals that had a tendency to vary would have the most advantage—hence groups would gradually become more ramified. Darwin described his theory of divergence in the *Origin,* pp. 111–26. For historical analyses of his theory, see Dov Ospovat, *The Development of Darwin's Theory* (Cambridge: Harvard University Press, 1981), pp. 170–83; and David Kohn, "Darwin's Principle of Divergence as Internal Dialogue," *Darwinian Heritage,* pp. 245–58.

FIGURE 15. Charles Darwin; photograph taken in 1860, just after publication of the *Origin of Species*.

others will have to be improved in a corresponding degree or they will be exterminated."[51] The general improvements created by natural selection, Darwin suggested elsewhere in the *Origin,* would not be merely relative to local environments. He appears to have thought natural selection would produce ever more progressive types, so that "the more recent forms must, on my theory, be higher than the more ancient; for each new species is formed by having had some advantage in the struggle for life over other and preceding forms."[52]

In the *Origin,* Darwin augmented the power of this progressive dynamic by attributing to natural selection the beneficent concern for the good of creatures, a concern that had been formerly expressed by the recently departed Deity. Artificial selection of animals was capricious and governed only by selfish desires of men; natural selection altruistically looked to the welfare of the creatures selected. As Darwin put it: "Man selects only for his own good; Nature only for that of the being which she tends."[53] He concluded that natural selection therefore insured progressive evolution: "as natural selection works solely by and for the good of each being, all corporeal and mental endowments will tend to progress towards perfection."[54]

Lyell, Darwin's friend and mentor, had decried the belief in geological progress because it represented an unwarranted theological intrusion into science; moreover, it was a conception dangerous to the unique moral character of human beings—since man was not merely an advanced orang. Many contemporary historians have shared Lyell's concern about the intrusion of alien wisdom into science, usually today understood as conservative political ideology. They have consequently abjured any interpretation of neo-Darwinism that appears to suggest progressive development, which seems to

51. Darwin, *Origin of Species,* p. 106.
52. Ibid., p. 337.
53. Ibid., p. 83.
54. Ibid., p. 489.

give succor to tainted political values. They imagine, however, that Darwin acquiesced in their understanding. But, as we have seen (and further evidence will be supplied in the next chapter), Darwin was not of a late-twentieth-century mind. Progress was the intended consequence of the model he had originally adopted for species evolution, namely, that of individual evolution. Indeed, the concepts of species change he had become familiar with from his grandfather, from Lamarck, from Grant, and, indirectly, from Tiedemann, Treviranus, Serres, and Geoffroy—all supposed progressive advance of creatures. Darwin worked a compromise in his conception of species transformation by making progress non-necessary, though general, and by placing the guide for advance in the external environment. This sort of compromise has perhaps deflected attention from the conception of embryogenesis that I believe stabilizes the core of Darwin's new theory of species evolution. We should now reconsider the gradual, uniform development of his embryological ideas and their role in his species theory.

5

DARWIN'S EMBRYOLOGICAL THEORY OF PROGRESSIVE EVOLUTION

Darwin formulated and reformulated his ideas of species change in light provided by the model of embryological evolution and recapitulation theory. The conceptual development of the problem of species change, particularly from the late 1820s through the 1840s, constrained him to regard recapitulation as a central part of the more general doctrine he would defend. The German recapitulationists and Green had advanced the ideas of individual evolution and species evolution as part of a common theory of progressive development of organic forms. Grant—whose advocacy of Lamarckism stuck to the memory of the elder Darwin as he composed his *Autobiography*—had made recapitulation theory an overt feature of his anatomical compositions while covertly orchestrating his transformationist ideas into public anatomical lectures. Lyell conceptually joined individual and species evolution in his analyses of Lamarck and Tiedemann. And Owen, whose views I will discuss more thoroughly in a moment, attacked both individual and species evolution as part of the same, unacceptable doctrine. Darwin thus inherited the two sorts of evolution as mutually implicative, formed them into a common conception, and defended them together against Lyell and especially Owen. The principle of recapitulation and the embryological model allowed Darwin to resolve the single most pressing problem presented to him by his professional colleagues: how to account for the unity of type, the Cuvierian *embranchements* of design, that almost every leading naturalist of the time recognized in the animal kingdom. This problem hovered over Darwin's early efforts at constructing his theory of evolution, and his deferred resolution at the end of the *Origin of Species* only disguises, to modern eyes, its critical role in the formation of the theory occupying the first twelve chapters. Darwin of course employed many other conceptual instruments in the construction of his theory; he worried about a myriad of problems—though none, I think, was quite so encompassing and significant for his general conception of the evolutionary process. By focusing on the principle of recapitulation and the embryological model—and the problems they were meant to resolve—Darwin's theory of evolution stands in a light of sepia tone, decidedly more in keeping with its early-nineteenth-century origins. This perspective will undoubtedly seem exaggeratedly antique to those historians and philosophers who have striven to distinguish Darwin's notion of contingently branching evolution from conceptions modeled on embryological development, of which those of Spencer and Haeckel represent the looming specters.

The Embryological Model as Formulated in the Notebooks

In his very earliest reflections on the subject, as recorded in his "Notebook B" during summer of 1837, Darwin modeled species evolution on embryogenesis. And the link by which he immediately joined the model of embryo development to the process of morphological change in species was the principle of recapitulation. Darwin was apparently inspired to articulate the process initially in this way by his grandfather's *Zoonomia*—in which embryogenesis also served as the model for species transmutation.[1]

Erasmus Darwin had sought simultaneously to combat Haller's embryological preformationism and Moses' species

1. Robert J. Richards, *Darwin and the Emergence of Evolutionary Theories of Mind and Behavior* (Chicago: University of Chicago Press, 1987), pp. 36–37.

FIGURE 16. Erasmus Darwin, 1731–1802, whose embryological model of evolution was adopted by his grandson Charles; portrait done in 1770.

preformationism. He marshaled evidence of metamorphosis in insects, imperfect and monstrous births, hybridization, and alterations of domestic animals—all to argue that the embryo must undergo gradual transformation during gestation. He believed, contrary to the older theory of evolution, that irritabilities and sensibilities of living matter led organisms (including the embryo) to adopt habits and modes of behavior that changed anatomical structures, so that the fetus would be gradually transformed in the womb. His embryological analysis had, he conceived, direct consequences for understanding species development:

> From this account of reproduction it appears, that all animals have a similar origin, viz. from a single living filament; and that the difference of their forms and qualities has arisen only from the different irritabilities and sensibilities, or voluntarities, or associabilities, of this original living filament . . . And that from hence, as Linnaeus has conjectured in respect to the vegetable world, it is not impossible, but the great variety of species of animals, which now tenant the earth, may have had their origin from the mixture of a few natural orders.[2]

Not only did the elder Darwin's blood pulse in the veins of his grandson, but his suggestions, ideas, and poetical fancies warmly surged through the mind of Charles. The younger Darwin's own mechanism of heritable modifications from habit probably derived, in part at least, from ruminations on his grandfather's work. As well, he likely elaborated Erasmus's considerations of a parallel development of the embryo and the species. Charles certainly pursued this line of thought in the early pages of that notebook, his first "Transmutation Notebook," which bears the same title as his grandfather's volume.

On the very first page of his "Zoonomia" notebook, for in-

2. Erasmus Darwin, *Zoonomia or the Laws of Organic Life,* 2d ed. (London: Johnson, 1796), 1: 502.

stance, Darwin mentioned two kinds of generation that his grandfather had distinguished, the "coeval" kind, that is, budding, splitting of planaria, and so on—in which the new product remained identical to the source—and sexual generation: "The ordinary kind ⟨the⟩ which is a longer process, the new individual passing through several stages (?typical, ⟨of the⟩ or shortened repetition of what the original molecule has done)."[3] Darwin suggested here that in sexual generation the new embryo passed through the "type" stages in a "shortened repetition," which the original living molecule (or "filament," to use his grandfather's term) had passed through in the transformation of species. Darwin thus opened his first notebook devoted to his species hypothesis with a formulation of the principle of recapitulation—a principle which achieved a synthesis between the older theory of evolution (preformationism) and Erasmus's theory of a parallel epigenetical development of embryo and species.

A bit further in this same "Zoonomia" notebook, Darwin considered that the goal, the final cause of sexual generation, was to provide in the neonate a new, original layer of the species by which it could adaptively respond to the environment. This was possible, he speculated shortly thereafter, because of the earlier layers of progressive adaptation preserved in embryonic development: "An originality is given (& power of adaptation) is given by true generation, through means of every step of progressive increase of organization being imitated in the womb, which has been passed through to form that species.—⟨Man is derived from Monad⟩."[4] Darwin did not simply move inertially with his grandfather's theory, but he extended and modified it, as this quotation indicates. He also pulled up new evidence that he chanced across in his

3. Charles Darwin, "Notebook B," MS p. 1, in *Charles Darwin's Notebooks, 1836–1844,* ed. Paul Barrett, Peter Gautrey, Sandra Herbert, David Kohn, and Sydney Smith (Ithaca: Cornell University Press, 1987), p. 170. Double wedge quotes indicate Darwin's insertions; single wedges indicate his deletions. This convention will be followed throughout the chapter.

4. Darwin, "Notebook B," MS p. 78 (Barrett et al., p. 190).

reading. He found confirmation for embryo-species parallelism, for instance, in an article by Serres which he read in December of 1837.[5] In the article, the French physiologist
declared that "mollusks are the permanent embryos of the
vertebrates and of man."[6] The following summer, in late August, Darwin again reiterated the idea that recapitulation—a
"condensation of changes"—would provide the individual
with a kind of platform upon which progressively to build adaptations. Presciently he began this notebook entry with a
conviction that would be realized shortly thereafter: "27th
August. There must be some law, that whatever organization
an animal has, it tends to multiply & IMPROVE on it.—
Articulate animals must articulate. in vertebrates tendency to
improve in intellect,—if generation is condensation of
changes, then animals must tend to improve."[7] A month later,
on 28 September, Darwin read Malthus and therewith discovered the law that he knew had to exist.

Natural selection, as I have indicated in the previous chapter, furnished precisely the force that could build new species
traits on the end of a progressive embryological evolution.
Later, in his Essays of 1842 and 1844 and in the *Origin of
Species,* Darwin would settle this principle of adaptation into
the foundation of his general theory of progressive evolution:
he would propose that heritable modifications occurred generally only at the end of individual development, that is, in the
more adult organism; such terminal additions would then be
recapitulated in the embryos of succeeding generations.[8] In

5. Darwin, "Notebook B," MS p. 163 (Barrett et al., p. 211).

6. Etienne Reynaud Serres, "Zoologie: Anatomie des mollusques," *L'Institut, section des Sciences mathématiques, physiques et naturelles* no. 191 (4
January 1837): 370.

7. Darwin, "Notebook D," MS p. 49 (Barrett et al., p. 347).

8. I will explore this principle of adaptation further below, where it will be
contrasted with Gould's alternate understanding of Darwin. Jonathan Hodge
has analyzed the full range of Darwin's early theorizing about generation and
its mechanisms. He discovers a deep continuity in Darwin's thought concerning these subjects, a continuity that stretches from Darwin's early notebook
jottings to his work in the 1860s. See M. J. S. Hodge, "Darwin as a Lifelong

this way Darwin's newly discovered "law" of natural selection functioned as that unifying force that earlier recapitulationists proposed as necessary to explain the parallel evolution of individual and species.

The principle of recapitulation, almost from the beginning of Darwin's species work, served as the aortic connection for three components of his early theory: the embryological model of evolution, the idea that the purpose of generation was progressive development, and the assumption of common descent. Even after he had formulated the law of natural selection, Darwin continued to pivot his theory of evolution around the principle of recapitulation, as this loosely flowing passage in December 1838 suggests:

> Seeing that ⟨Man⟩ ⟨⟨all vertebrates . . . ⟩⟩ can be traced to a germ, endowed with the vital principle . . . & knowing from analogy, that all these very animals are descended from some one single stock,—one is led to suspect that the birth of the species & individuals in their present forms, are closely related—by birth the successive modifications of structure being added to the germ, at a time, (as even in childhood) when the organization is pliable, such modifications, becomes as much fixed, as if added to old individuals, during thousands of centuries, each of us, then ⟨is as old, as the oldest animal⟩, have passed through as many changes, as has any species.[9]

In the most creative and heated phase of his theory construction, Darwin supposed that each of us during embryogenesis comes to "pass through" the evolutionary history of our species. Thus at the core of his emerging theory of descent, which stands firmly fixed upon an embryological model of

---

Generation Theorist," in *The Darwinian Heritage,* ed. David Kohn (Princeton: Princeton University Press, 1985), pp. 207–43. We should not be surprised, therefore, if Darwin's early notions about recapitulation had become deeply embedded within his maturing theory of species change.

9. Darwin, "Notebook E," MS pp. 83–84 (Barrett et al., p. 418).

species change, and about which are layered conceptions of adaptation through terminal additions, progressive advance in the organic improvement of life, and natural selection as the law governing these processes—at the core of these ideas lies the venerable thesis of recapitulation.

Owen's Rejection of Recapitulation and Evolution

Darwin's early reflections on recapitulation stumbled against the authority of Richard Owen. Owen's comprehensive knowledge of anatomy had been hard won. Because his family had been financially strapped, he had to make his own way, first as a midshipman in the Royal Navy, then as an indentured apprentice to two surgeons, finally reaching medical school at Edinburgh, arriving the year (1824–25) before Darwin came up. He apparently made the acquaintance of Robert Grant,[10] who perhaps spoke of the enthusiasm that occupied him at the time, namely, Lamarck's theory of transmutation. While attracted by such attendant intellectual diversions, Owen yet found the formal curriculum at Edinburgh sclerotic. He came down from the school at the end of his second term, next taking an apprenticeship at Saint Bartholomew's Hospital in London. After obtaining a license in 1826, he began a small practice around Lincoln's Inn Field, chiefly among lawyers, which must have inclined him to think of other employment. The following year he started in a minor position at the Royal College of Surgeons that would initiate his scientific career; he was to assist in the preparation of a catalog of the anatomical collection that John Hunter had bequeathed the College. While working on this project, he came under the tutelage of Joseph Henry Green. Owen's extraordinary industry and talent won out. He became the first Hunterian Professor at the College in 1836 and delivered annually the Hunterian lectures during his tenure. Owen's reputation has

10. See Phillip Sloan's introduction to *Richard Owen's Hunterian Lectures, May–June 1837* (London: British Museum of Natural History; Chicago: University of Chicago Press, forthcoming).

suffered because of a difficult personality, which did not age well. In later years, his talents with the dissecting scalpel could not match Huxley's talents with the incisive barb. His relations with Darwin began cordially as the two worked on the large number of specimens returned from the *Beagle* voyage. But when Owen later wrote an anonymous review of the *Origin of Species* and claimed that the theory of evolution was unsupportable but that in any case Richard Owen had first conceived it[11]—then he sealed his fate not only for the Darwinians, who easily guessed his authorship, but also for recent historians.

Owen had initially given some approbation to the principle of recapitulation.[12] His casual endorsement may have come because Hunter had some claim for originally devising the idea. Moreover, his mentor Green had endorsed recapitulation and cited Hunter as having first formulated the principle and Tiedemann as having demonstrated it.[13] Yet in spring of 1837, Owen turned away from the thesis. Darwin, annotating his notebook paraphrase of Serres's mollusk remark, added "Owen says Nonsense."[14] Darwin probably discussed the principle personally with Owen, though he may have simply recalled Owen's remarks in the Hunterian Lecture of 9 May 1837.

Owen's task in that lecture was to discuss the work and anatomical collections of John Hunter, and this he did through-

11. [Richard Owen], "Darwin on the Origin of Species," *Edinburgh Review* 11 (1860): 487–532.

12. Richard Owen, "Remarks on the Entozoa," *Transactions of the Zoological Society of London* 1 (1835): 387–94. The remark (on p. 390) would appear to endorse the principle of recapitulation: "all the classes of the *Acrite* division exhibit the lowest stages of animal organization, and are analogous to the earliest conditions of the higher classes." Darwin read this as Owen's avowal of recapitulation.

13. See n. 3, chap. 3, and n. 23, chap. 4.

14. Darwin, "Notebook B," MS p. 163 (Barrett et al., p. 211). Darwin later delighted to observe that Owen had in fact endorsed the principle of recapitulation as recently as the 1835 article on the Entozoa. See n. 12 and Darwin, "Notebook C," MS p. 48 (Barrett et al., p. 254).

FIGURE 17. Richard Owen, 1804–1892, the Hunterian Professor armed for battle against the evolutionists; engraving. Wellcome Institute Library, London.

out most of his presentation. He noted that "Hunter not only points out the analogy of the transitory stages of the Embryo bird to the permanent structure of less complicated Animals but applies the conditions to the explanation of Monstrosities."[15] Hunter had suggested, using his version of recapitulation, that monsters were animals which had not completed their embryonic transformation to a final adult form. Like Geoffroy Saint-Hilaire, Hunter regarded a monster as arrested at an earlier developmental stage, one similar to the adult form of a lower species. Owen simply and sympathetically laid out Hunter's position, attempting to distinguish it a bit from Geoffroy's but certainly not rejecting it. (Indeed, in the 1860s it became the basis for his own, post-Darwinian theory of evolution.)[16] But at the end of that lecture of 9 May, he seemed to pirouette in an opposite direction.

Owen considered the "philosophic inquiry" of men like "Tiedemann, Purkinge, Baer, Rathke, Wagner, Valentin, Müller," as having laid "the foundations of a just and true theory of animal development and organic affinities."[17] These thinkers had to be distinguished from the a priori school of transcendentalists, who had misrepresented "the beautiful observation of the resemblance of imperfect conditions of the organs of a higher species to the perfect conditions of corresponding organs in a lower organized species." Owen now expressly denied that "the Human Embryo repeats in its development that structure of any part of another animal; or that it passes through the forms of the lower classes." He linked his rejection of recapitulation to that of an even more pernicious idea: "The doctrine of transmutation of forms during

15. Richard Owen, in *Richard Owen's Hunterian Lectures,* lecture for 9 May 1837, MS p. 83 (Sloan, p. 61).

16. For an admirable discussion of the development of Owen's evolutionary ideas, see Evelleen Richards, "A Question of Property Rights: Richard Owen's Evolutionism Reassessed," *British Journal for the History of Science* 20 (1987): 129–71. See also n. 28, below.

17. Owen, in *Richard Owen's Hunterian Lectures,* MS p. 97 (Sloan, p. 67).

the Embryonal phases, is closely allied to that still more objectionable one, the transmutation of Species."[18]

The timing of Owen's lecture and internal evidence suggest that he came to reject recapitulation because he had just read, as he was finishing his lecture manuscript, Martin Barry's recently published account of von Baer's *Entwickelungsgeschichte der Thiere*.[19] In his two-part article, Barry cited all of the German physiologists whom Owen mentioned and carefully set out von Baer's theory of development and the several objections to recapitulation,[20] which objections Owen rehearsed at the end of his lecture. Barry also confirmed—quoting Gabriel Valentin's von Baerean line—that "the development of the animal kingdom, and of the individual animal, are in the original idea, throughout, one and the same."[21] Owen obviously appreciated the power of von Baer's animadversions on recapitulation and the connection of that principle with species transformation.

Owen reiterated his opposition to recapitulation theory in his Hunterian lectures of 1843 and 1844. He maintained, conformable to the opinions of Barry and von Baer, that the more perfect animals in their fetal development resembled the permanent forms of lower animals—but only at the earliest periods of embryogenesis and at the lowest levels of the species hierarchy. He cautioned that the "extent to which the resemblance, expressed by the term 'Unity of Organisation,' may be traced between the higher and lower organised animals, bears an inverse ratio to their approximation to matu-

18. Ibid., MS pp. 98–99 (Sloan, p. 68).

19. This is Phillip Sloan's persuasive argument. He bases his conclusion on the shift in discussion and on the physical differences in the last several pages of the manuscript.

20. Martin Barry, "On the Unity of Structure in the Animal Kingdom," *Edinburgh New Philosophical Journal* 22 (1836–37), pp. 116–41; and "Further Observations on the Unity of Structure in the Animal Kingdom," *Edinburgh New Philosophical Journal* 22 (1836–37), pp. 345–64. The articles appeared in the numbers for January and April 1837.

21. Barry, "On the Unity of Structure in the Animal Kingdom," p. 139.

rity."[22] Thus, "all mollusks are at one period like Monads, at another Acephalans [the lowest class of mollusk]; but scarcely any typify the Polypes, and none the Acalephes [lower and higher classes of radiata]."[23] Moreover, embryonic spiders do not metamorphose at all; rather "the mature form is sketched out from the beginning," much, he might have added, as Swammerdam believed.[24] From these analyses of embryological development, Owen concluded his 1843 lectures with the observation that such facts as he recounted "would of [themselves] have disproved the theory of evolution, if other observations of the phenomena of development had not long since rendered that once favourite doctrine untenable."[25] By "evolution" he meant the comprehensive process that supposedly produced embryological recapitulation and the parallel transformation of species.

In his lectures of 1844, Owen began where he had left off, namely, with the claim that the "hypothesis of the development of species by progressive transmutation" failed on the evidence: vertebrates only resembled invertebrates at the very earliest stages of development; embryology thus fore-

22. Richard Owen, *Lectures on the Comparative Anatomy and Physiology of the Invertebrate Animals, Delivered at the Royal College of Surgeons in 1843,* notes taken by William White Cooper and revised by Richard Owen (London: Longman, Brown, Green, and Longmans, 1843), p. 367. Owen obviously owed an unacknowledged debt to von Baer and Barry for the principle of ontogenetic divergence. He had been prepared, however, to accept their analyses by reason of his training with Green and his reading of Carus. Green had insisted that the developmental sequence of species was ramified and not lineal, and he ascribed the latter trajectory to Lamarck's theory (see my discussion above). Carus also urged that the morphological patterns of development in both species and fetus were divergently branching. He made this representation in the textbook, the *Lehrbuch der Zootomie,* that Green used in his courses and Owen studied during his apprenticeship. See Carl Gustav Carus, *Lehrbuch der Zootomie* (Leipzig: Gerhard Fleischer the Younger, 1818), pp. 667–70; see also nn. 8 and 77 of chap. 3 for relevant quotations.

23. Owen, *Lectures on the Comparative Anatomy and Physiology of the Invertebrate Animals* p. 368.

24. Ibid., p. 250.

25. Ibid., pp. 368–69.

shadowed no general unity of organization.[26] According to Owen's understanding of the Germanic and Lamarckian theories of evolution, the newborn individual was presumed to acquire species-transforming traits that would accumulate, forming the stages of transition through which the embryos of succeeding generations would pass. For Owen, evolution of species and evolution of embryo expressed the same idea; so to undermine one was to watch the other collapse.

Owen's authority stood over most of the subjects upon which Darwin worked during the period prior to the publication of the *Origin of Species,* and his opposition to the twin doctrines of recapitulation and species transformation formed the treacherous channel through which the younger naturalist had carefully to steer. For just as Owen regarded recapitulation and species transformation as virtually the same idea, which had to be opposed, Darwin thought them virtually the same idea, which had to be defended. Darwin continued to explore the principle of recapitulation throughout his species notebooks.[27] He more carefully formulated its significance, however, in his Essays of 1842 and 1844, and in the *Origin,* where he explicitly attempted to answer Owen and to adapt von Baer's view to his own recapitulation purposes.

In the Essays and *Origin,* Darwin's reflections on embryological recapitulation occur within his larger considerations of the doctrine of unity of types. He argued that morphological similarity of adults across, say, vertebrate species and their embryological similarity through ontogenetic development could both be explained most naturally by descent from a common ancestor rather than by divine creation

26. Richard Owen, *Lectures on the Comparative Anatomy and Physiology of the Vertebrate Animals, Delivered at the Royal College of Surgeons of England in 1844 and 1846* (London: Longman, Brown, Green, and Longmans, 1846), pp. 10–11.

27. Darwin, "Notebook B," MS pp. 78, 163 (Barrett et al., pp. 190, 211); "Notebook C," MS pp. 48–49, 149, 162 (Barrett et al., pp. 254, 284, 289); "Notebook D," MS pp. 170, 179 (Barrett et al., pp. 387, 390–91); and "Notebook E," MS p. 89 (Barrett et al., pp. 420).

according to a common ideal or archetype. This is the significance of the lines he jotted on the back flyleaf of his copy of
Owen's *On the Nature of Limbs* (1849): "I look at Owen's
Archetypes as more than idea, as a real representation as far
as the most consummate skill & loftiest generalization can
represent the parent form of the Vertebrata."[28] The generalized vertebrate archetype, Darwin suggested, was actually a
natural creation, a primitive organism whose descendants
had become specialized and differentiated through evolutionary adaptations. Those descendants would thus constitute the myriad of families, genera, and species united
within the vertebrate archetype. The spare archetypal pattern
of traits common to all vertebrates—essentially, according to
Owen, only a backbone with riblike processes (see fig. 25)—
would provide, Darwin thought, a fairly accurate picture of

28. Darwin's pencil annotation occurs on the back flyleaf of his copy of
Richard Owen, *On the Nature of Limbs* (London: Van Voorst, 1849). The book
is part of the Darwin Library, held in the Manuscript Room of Cambridge University Library. Darwin added one more sentence in annotation: "I followed
him that there is a created archetype, the parent of its class." This last remark
indicates Darwin recognized that Owen also harbored a naturalistic conception that edged toward his own; at least the last paragraph of Owen's book
(which Darwin heavily scored) suggested this (p. 86): "To what natural laws
or secondary causes the orderly succession and progression of such organic
phaenomena may have been committed we as yet are ignorant. But if, without derogation of the Divine power, we may conceive the existence of such
ministers, and personify them by the term 'Nature,' we learn from the past
history of our globe that she has advanced with slow and stately steps, guided
by the archetypal light, amidst the wreck of worlds, from the first embodiment of the Vertebrate idea under its old Ichthyic vestment, until it became
arrayed in the glorious garb of the Human form." With such passages in
mind, Michael Ruse suggests that Owen's claim to being an early species evolutionist prior to Darwin's publication had some faded justice. See his
Darwinian Revolution (Chicago: University of Chicago Press, 1979), p. 228.
Evelleen Richards thoroughly investigates Owen's claim, partly sustaining it,
in "A Question of Property Rights: Richard Owen's Evolutionism Reassessed."
She portrays Owen as flirting with a naturalistic account of species, only to
have his hand slapped by Adam Sedgwick and some conservative guardians
of religious orthodoxy. She believes that Owen then ran up his orthodox
colors, which he tried to strike after Darwin's success.

the Urancestor from which all the vertebrates descended. The embryo of a higher vertebrate might then, during embryogenesis, preserve a series of such pictures of its evolutionary forbears. In working out these ideas, however, Darwin had to first consider the objections to recapitulation advanced by von Baer and his followers.

Darwin's Knowledge of von Baer

Darwin did not mention von Baer by name in the Essays of 1842 and 1844, though he did in the *Origin*. The overt linkage of embryological ideas with the name of von Baer came only after he read Huxley's translation, done in 1853, of selections of the *Entwickelungsgeschichte*.[29] It is nonetheless clear in the Essays that the objections to recapitulation, which he there considered, were von Baer's. Darwin had four possible sources for learning of von Baer's treatment, one of which can be positively confirmed. The first likely source is Owen, who in his Hunterian lectures of 1837, 1843, and 1844 mentioned von Baer and reiterated several of the great embryologist's objections, though not always carefully attaching von Baer's name to them—indeed, at times seeming to take credit for them.[30] In his 1844 Essay, Darwin noted Owen's denial of the principle of recapitulation and of course may have discussed with him the original author of many of the objections.[31] A second possible source is William Carpenter's *Principles of General and Comparative Physiology*, the first edition of which appeared in 1839 and the second in 1841. Carpenter granted that "if we watch the progress of evolution, we may

29. Karl Ernst von Baer, "Fragments relating to Philosophical Zoology: Selections from the Works of K. E. von Baer," trans. Thomas Henry Huxley, in *Scientific Memoirs, Selected from the Transactions of Foreign Academies of Science, and from Foreign Journals: Natural History*, ed. Arthur Henfrey and Thomas Henry Huxley (London: Taylor and Francis, 1853), pp. 176–238.

30. Evelleen Richards traces Owen's effort to appropriate von Baer's thesis as his own. See "A Question of Property Rights: Richard Owen's Evolutionism Reassessed."

31. Darwin, "Essay of 1844," in *Foundations of the Origin of Species,* ed. Francis Darwin (Cambridge: Cambridge University Press, 1909), p. 219.

trace a correspondence between that of the germ in its advance towards maturity, and that exhibited by the permanent conditions of the races occupying different parts of the ascending scale of creation."[32] Here again the term "evolution" was used to refer indifferently to embryological recapitulation and species relationships, though in the first two editions Carpenter had in mind only "ideal," not historical, species relationships.[33] Further, Carpenter thought similarity of the two evolutions was vague and really merely a consequence of von Baer's "law," namely, that "a heterogeneous or special structure arises out of one more homogeneous or general; and this by a gradual change."[34] He then repeated von Baer's basic objections to strict recapitulation. Darwin's reading notebooks indicate he examined the fourth edition (1854) of Carpenter's volume.[35] Likely he also read either the first or second editions, but there is no direct evidence. Concerning a third source, we know that Darwin studied Johannes Müller's account of von Baer's work in the newly translated second volume (1842) of *Elements of Physiology*.[36] Darwin's

32. William Carpenter, *Principles of General and Comparative Physiology* (London: Churchill, 1839), p. 170.

33. In the fourth edition (1854) of Carpenter's book, he did consider the doctrine of "an actual 'transmutation' of the lower forms into the higher [taking] place in the course of geological time; so that, from the germs first introduced, or from others which have since originated in combinations of inorganic matter, the whole succession of organic forms, from the simplest Protophyte up to the Oak or Palm, from the Protozoon up to Man, has been gradually evolved." Carpenter was reacting specifically to the doctrine of Robert Chambers's anonymously published *Vestiges of the Natural History of Creation,* which in its turn had depended in part on Carpenter's account of recapitulation theory in the earlier editions of his book. See William Carpenter, *Principles of Comparative Physiology,* 4th ed. (London: Churchill, 1854), p. 106.

34. Carpenter, *Principles of General and Comparative Physiology* (1st ed., 1839), p. 170.

35. Darwin's copy of the fourth edition of Carpenter's book is preserved in the Manuscript Room at Cambridge University Library.

36. Johannes Müller, *Elements of Physiology,* trans. William Baly (London: Taylor & Walton, 1837–42), 2: 1591–92. Darwin's "Reading Notebooks" indicate he examined the second volume of Müller's *Elements* in early April 1842. The Darwin Library at Cambridge retains his copy of Müller, with the second

1842 Essay shows the impact of having just read Müller; and the 1844 Essay explicitly refers to Müller's objections, which indeed were only Müller's report of von Baer's own. The most interesting possibility, though, is the initial source for Owen's and Carpenter's knowledge, namely, the two memoirs of the Scots physician Martin Barry, "On the Unity of Structure in the Animal Kingdom," which appeared in the numbers of the *Edinburgh New Philosophical Journal* for January and April 1837.[37] Darwin frequently met with Owen during this period of preparation of his *Beagle* materials, and the two men seem to have discussed a range of issues of common interest. So Owen may well have drawn Darwin's attention to Barry's articles. Moreover, Darwin kept up with the Edinburgh journal, and from 1839 on had a subscription.

The second of Barry's memoirs, which was almost entirely devoted to von Baer, appeared in April 1837, about the time (within a month or so) that Darwin opened "Notebook B," his first transmutation notebook. What makes this conjunction so tantalizing is that in explicating von Baer's notion of types and the degrees of their complex formation, Barry illustrated it by what he called "The Tree of Animal Development."[38] This tree had as its root the germ, which was morphologically similar in all animals; and from this common ancestral monad branched off the invertebrates on one side and on the other the vertebrates, with the fishes, mammals, and finally man ramifying through progressively higher branches of the tree (see fig. 18). Barry diagrammatically sketched, without labels, the three main stems of the invertebrates, each coming directly off the germ; and his more elaborate depiction of three types of vertebrates also included elliptical representations of the branching of reptiles and birds. In the ascent of

volume well scored on p. 1592, where von Baer is discussed. See also Charles Darwin, "Reading Notebooks," in appendix 4 in *Correspondence of Charles Darwin,* 4: 464.

37. See n. 20, above, for bibliographic information.

38. Barry, "Further Observations on the Unity of Structure in the Animal Kingdom," p. 346.

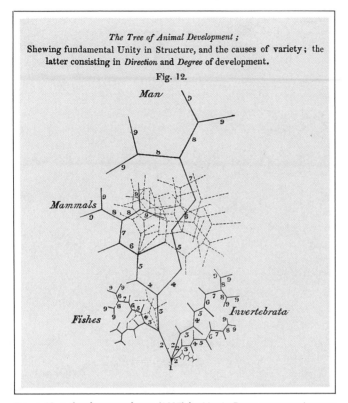

FIGURE 18. A developmental tree (1837) by Martin Barry, representing von Baer's conception of the vertebrate and invertebrate archetypes and their developmental patterns.

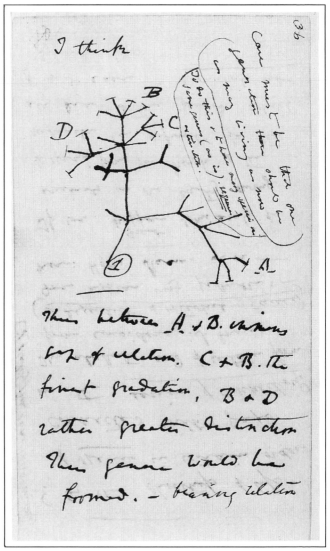

FIGURE 19. A descent tree (1837) from Darwin's "Notebook B" depicting species branching, with genera (at the nodes) also indicating the ancestors. "Thus genera would be formed.—bearing relation to ancient types."

the branches, the nodes represented the more generalized classes, orders, families, and genera, with the branch ends the more particularized species, varieties, and individual characters.

If Barry's illustration were viewed as he meant it to be, it offered a nice representation of von Baer's theory of the four archetypes and their respective developmental formations. However, if viewed through Darwinian eyes, the "Tree of Animal Development" becomes transmogrified. Indeed, Barry's illustration resembles Darwin's famous sketch, on page 36 of "Notebook B," of a descent tree—even to both sketches beginning with a simple monad or germ labeled "1" (see fig. 19). Under his sketch Darwin observed "Thus genera would be formed—bearing relation to ancient types." That is, the nodes representing the generic patterns would also represent the morphological types of the ancestors. There is no direct evidence that Darwin took his inspiration from Barry, but the puzzle pieces fit snugly together; they look homologous. In any case, Darwin's annotation in Owen's book about the archetype being the ancestor supplies the key to his conversion of von Baer's law back into the principle of recapitulation. Darwin worked this transformation from the Essays, through subsequent studies, and into the *Origin.*

Historical Evaluation of Darwin's Principle of Recapitulation

I will attempt to demonstrate Darwin's transformation of von Baer in a moment. But let me here quickly encapsulate the assessments made by historians of Darwin's view on recapitulation and of its implications for his theory of evolution. The general opinion that Darwin rejected recapitulation was first moved by E. S. Russell in his extremely influential book *Form and Function* (1916). Without much analysis, Russell simply asserted that Darwin "avoided the snare of the Meckel-Serres theory of recapitulation, according to which the embryo of the highest animal, man, during its development climbs the ladder upon the rungs of which the whole animal

series is distributed, in its gradual progression from sim-
plicity to complexity." Russell found his own hero as the
source of Darwin's embryology: "The law of development
which he adopts is that of von Baer, which states that development is essentially differentiation, and that as a result embryos belonging to the same group resemble one another the
more the less advanced they are in development."[39] Most
historians have followed Russell, with the marked excep-
tion of Jane Oppenheimer. In a foundational article (1959),
Oppenheimer examined Darwin's Essays and the *Origin* in
light of von Baer's theory. Unlike Russell, she concluded that
Darwin failed to heed the wisdom of his German predeces-
sor: he endorsed recapitulation and thereby prepared the
way for the gross mischief that Haeckel's biogenetic law
worked on three-quarters of a century of embryology.[40] In his
comprehensive *Ontogeny and Phylogeny* (1977), Stephen
Jay Gould rescued Darwin, bringing him back on board with
von Baer—but happily kept Haeckel submerged in the mael-
strom of chaotic and dangerous theory. Gould maintained
that "Darwin had accepted the observations of von Baer—a
flat denial of recapitulation and its obvious evolutionary
meaning."[41] Gould argued that in the *Origin* Darwin appeared to come close to recapitulation, but actually he
only repeated von Baer's thesis that the embryo developed
through more general forms to more particular ones. There
would thus be resemblances among embryos of different
species at earlier stages of development, but the embryo did
not repeat the *adult* forms of lower species.[42] Ernst Mayr,

39. E. S. Russell, *Form and Function: A Contribution to the History of Animal Morphology* (London: Murray, 1916; reprint, Chicago: University of
Chicago Press, 1982), p. 236.

40. Jane Oppenheimer, "An Embryological Enigma in the *Origin of Species,"* in *Forerunners of Darwin,* ed. Bentley Glass (Baltimore: Johns Hopkins
University Press, 1968), pp. 292–322.

41. Stephen Jay Gould, *Ontogeny and Phylogeny* (Cambridge: Harvard
University Press, 1977), p. 70.

42. Ibid., p. 70.

though a bit more cautious than Gould, yet concluded in his *Growth of Biological Thought* (1982) that "most authors (including Darwin) rejected the claim that ontogeny is the recapitulation of the *adult stages* of the ancestors."[43] Dov Ospovat provided the most searching and rich analysis of Darwin's use of recapitulation. In illuminating fashion he discussed Darwin's relation to both von Baer and Owen on these matters; and his conclusions about the Darwin of the Notebooks do not differ markedly from my own. However, Ospovat claimed that "Darwin abandoned recapitulation when it began to lose favor [in the 1840s] among younger biologists whose opinions he respected."[44] In regard to the Essays and the *Origin,* he thought that "it is safe to place Darwin among the followers of von Baer, rather than of the recapitulationists."[45]

Peter Bowler, in his recent *Non-Darwinian Revolution* (1988), captures much of the common sentiment. He argues that the developmental model of species evolution prevailed throughout the nineteenth century—except with Darwin. The "belief that ontogeny (individual growth) recapitulates phylogeny (the history of the type) thus owes little to Darwinism, and is more characteristic of the non-Darwinian, or developmental, view of evolution."[46] It was rather "Haeckel, not Darwin, who popularized the recapitulation theory as an integral part of late-nineteenth-century evolutionism." It was Haeckel, not Darwin, who drew from the developmental model "a progressionist vision of evolution with a main line of development aimed at mankind."[47] In Bowler's view, the essence of Darwinism is natural selection, which crushes

43. Ernst Mayr, *The Growth of Biological Thought* (Cambridge: Harvard University Press, 1982), p. 475.

44. Dov Ospovat, *The Development of Darwin's Theory* (Cambridge: Harvard University Press, 1981), p. 152.

45. Ibid., p. 162.

46. Peter Bowler, *The Non-Darwinian Revolution: Reinterpreting a Historical Myth* (Baltimore: Johns Hopkins University Press, 1988), p. 11.

47. Ibid., p. 13.

recapitulation and devours any notions of evolutionary progress.

In the previous chapter and elsewhere,[48] I have tried to show that Darwin's theory both before and after Malthus remained a theory of evolutionary progress.[49] And I have argued in this chapter thus far, and will continue in what follows, that the rib around which Darwin fleshed his idea of progressive development was the principle of recapitulation. By contrast to this view, I believe we see in the history of recent historiography Darwin gradually becoming a neo-Darwinian, a modern biologist who scorns notions of progress and rejects German transcendental theories.[50] But to

48. See Robert J. Richards, "Moral Foundations of the Idea of Evolutionary Progress," in *Evolutionary Progress,* ed. Mathew Nitecki (Chicago: University of Chicago Press, 1988).

49. Dov Ospovat, John Greene, and Michael Ruse have also detected the deeply embedded progressive streaks in Darwin's theory. See Ospovat, *The Development of Darwin's Theory,* pp. 210–28; John Greene, *Science, Ideology, and World View* (Berkeley: University of California Press, 1981), pp. 128–57; and Michael Ruse, "Molecules to Men: Evolutionary Biology and Thoughts of Progress," in Nitecki, *Evolutionary Progress.*

50. In addition to Bowler's opinions about Darwin's theory as antiprogressivist, Ernst Mayr and Stephen Jay Gould have also stressed that Darwin's theory does not permit progress in evolution. Mayr maintains that "Darwin, fully aware of the unpredictable and opportunistic aspects of evolution, merely denied the existence of a lawlike progression from 'less perfect to more perfect.'" See Ernst Mayr, *Growth of Biological Thought,* p. 531. Gould has argued that "to Darwin, improved meant only 'better designed for an immediate, local environment.'" This is because natural selection "proposes no perfecting principles, no guarantee of general improvement; in short, no reason for general approbation in a political climate favoring innate progress in nature." See Stephen Jay Gould, *Ever Since Darwin* (New York: Norton, 1977), p. 45. More recently, Gould has detected some tension in Darwin's own understanding. "Darwin," he asserts, "identified this denial of general progress in favor of local adjustment as the most radical feature of his theory." Yet Darwin, Gould recognizes, sometimes certainly sounded like a progressionist. Gould resolves this as a personal conflict within Darwin's own psyche, not as a matter of his theory: "The logic of the theory pulled in one direction, social preconceptions in the other. Darwin felt allegiance to both, and never resolved this dilemma into personal consistency." See Stephen Jay Gould, *Wonderful Life: The Burgess Shale and the Nature of History* (New

catalyze this observation, let me turn now to the Darwin of the Essays, the Barnacle books, the *Origin,* and the *Descent of Man.*

Recapitulation in the Essays and the Impact of Agassiz's Fishes, 1842–1844

In the 1842 Essay, Darwin situates his discussion of embryology within the wider context of the newly named "science of 'Morphology,'" which concerns itself with the problems of unity of type. Within this perspective, the "soberest physiologists" (apparently convinced by the transcendentally intoxicated Goethe) admit that the plant consists of metamorphosed leaves and that the "skulls of the vertebrates are undoubtedly composed of three metamorphosed vertebrae." In a similar fashion, the wing of a bird, the paddle of a porpoise, the hoof of a horse, and the hand of a man exhibit the same unity of design. These transcendental relationships could be explained, Darwin confidently argues, on the basis of his theory, which ascribes morphological unity to descent from a common ancestor. And since "this general unity of type in great groups of organisms . . . displays itself in a most striking manner in the stages through which the foetus passes," embryological recapitulation ought also to fall within the explanatory purview of the theory of descent.[51]

York: Norton, 1989), pp. 257–58. As I have tried to argue above, and I will provide further evidence below, Darwin's progressivism was not merely a personal fancy, beguiled by the times; it was deeply embedded in the structure of his theory.

51. I have compared the published version of the 1842 and 1844 Essays with the originals preserved in the Darwin archives. The published essays are in Francis Darwin, ed., *The Foundations of the Origin of Species;* see n. 31, above. The quotation comes from p. 42. The original 1842 Essay is in DAR 6; and the rough draft of the 1844 Essay is in DAR 7. (The standard DAR numbering for the Darwin manuscripts will henceforth be used in my notes.) Darwin had a fair copy made of his draft of the 1844 Essay, and this was the version Francis Darwin published. Darwin's knowledge of Goethe was filtered through William Whewell, in his *History of the Inductive Sciences* (London: Parker, 1837), 3: 433–48.

In the very early stages of embryological development, Darwin points out, the arterial systems of fish, bird, and mammal cannot be distinguished. But then he immediately makes the objection: "It is not true that one passes through the forms of a lower group, though no doubt fish more nearly related to fetal state." This objection, often cited by historians, was, however, short-lived, likely born of his recent examination of Johannes Müller's account of von Baer's theory.[52] In the middle of the manuscript page, jotted perhaps when revising this sketch for his more elaborate 1844 Essay, he pens and then circles: "They pass through the same phases, but some, generally called the higher groups, are further metamorphosed."[53] He seems to mean that the embryo does indeed preserve the earlier types, but that species of those types have further evolved since ancient times. In short, the embryo does not go through the adult stages of lower creatures that *exist in the present day*. Darwin concludes the 1842 Essay by indicating why embryos preserve earlier forms; namely, because of the homogeneity of the maternal environment, "there is no object gained in varying form &c. of foetus."[54] Natural selection will tend to produce alternations only after the fetus pops from the womb.

Darwin's confidence in recapitulation theory, which momentarily waned in early 1842, waxed during the next two years, largely as the result of the influence of the great Swiss naturalist Louis Agassiz (1807–73), who throughout his life remained an aggressive foe (defanged by the late 1860s) of transmutation theory. Darwin began reading Agassiz's studies of glaciation in the last part of the 1830s and opened a correspondence with him on that subject in March 1841.[55]

---

52. Darwin read Müller in April 1842. He wrote the 1842 Essay during May and June of that year.

53. Darwin, "1842 Essay," in DAR 6, MS p. 42.

54. Darwin, "Essay of 1842," *Foundations of the Origin of Species,* pp. 43–45.

55. Charles Darwin to Louis Agassiz (1 March 1841), *Correspondence of Charles Darwin,* 2: 284.

great number of species whose remains are known to us, the evidence seems conclusive that all the Fishes of that time, whatever may have been their degree of development in other respects, could have not advanced beyond the embryonic grade of the greater number of existing Fishes, as regards the structure of their spinal column.   Moreover, in nearly all the earlier Fishes, as was first pointed-out by Prof. Agassiz, we find a conformation of the tail which differs from that prevailing amongst the existing Fishes, but corresponds with that which presents itself in the embryonic state of the latter.   For in most of the Osseous fishes of the present epoch, the bodies of several of the terminal caudal vertebræ coalesce, so that the spinal column appears to end abruptly, whilst their neural and . hæmal arches and spines are equally developed above and below, so as to form the 'homocercal' tail represented in Fig. 78, A; in almost every Fish anterior to the Liassic period, on the other hand, the tail was formed upon the 'heterocercal' type, the vertebral column being continued onwards into its upper lobe, which is consequently the largest (B).

Fig. 78.

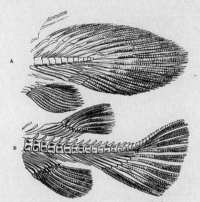

A, Homocercal tail ; B, Heterocercal tail.

Now it is obiously the 'heterocercal' tail, which departs least from the 'archetype;' and we find that even those Fishes which present the 'homocercal' conformation in their mature condition, have their tails originally 'heterocercal.'   Thus as the 'heterocercal' tail is the *most general* character of the class, being possessed by every fish at some period of its existence, whilst the 'homocercal' conformation is specially limited to a section of the class, the all-but-universal prevalence of the former during the earlier periods of the life of the class in our seas, and the comparatively late appearance of the latter, constitute a very remarkable example of this form of the doctrine above stated.—The Geological history of the

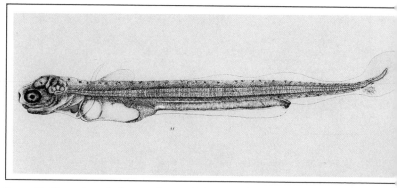

FIGURE 21. Illustration of salmon embryo with heterocercal tail, from Louis Agassiz and Carl *Histoire naturelle des poissons d'eau douce* (1842–1845).

Darwin's initial interest in this topic came from his effort to explain the terracelike "parallel roads" of Glen Roy. In the mid-1840s, though, it was fossil and embryonic fish that fixed his attention. In 1842, Agassiz, in a continuation of a massive study, published with his assistant Carl Vogt three volumes on fossil and living fish. In the second volume—on the embryology of salmon—Vogt reported that the tail of the fetal fry went through a stage of development parallel to that found by Agassiz in the fossils of the oldest fish.[56] In the adult salmon,

56. Carl Vogt, *Embryologie des salmones,* vol. 2 of Louis Agassiz, *Histoire naturelle des poissons d'eau douce* (Neuchâtel: Petitpierre, 1842). Vogt's discussion occurs on pp. 256–57.

URE 22. Fossils showing ganoid fish with heterocercal tails, from Louis Agassiz, *Monographie des ssons fossiles du vieux grès rouge, ou système Devonien* (1844–1845).

the spinal cord ended in a small symmetrical splay of its neural and hemal arches—what Agassiz called the homocercal
tail (see fig. 20). But in an early stage of embryonic development, the spinal cord arched upward into the superior lobe
of the tail fin (see fig. 21)—just as in the heterocercal tail of
fossil fish (see figs. 20 and 22). Richard Owen mentioned this
discovery almost immediately in his essay "On British Fossil
Reptiles," published in the *Edinburgh New Philosophical*

Journal in April 1842.[57] Darwin read this article and marked it up. In the same journal during the next year, he examined Agassiz's own brief account and scored the lines:

> it is remarkable that all the fishes of the greywacke [i.e., transitional] and coal formations belong to the Placoidians (Rays and Sharks), and to the strange looking Ganoidians, whose vertebral column, bent upwards in the caudal fin, presents a configuration impressed upon the full-grown fish which, at the present day, amongst bony fishes, is found only in embryos.[58]

Agassiz believed that numerous divine creations and extinctions within the *embranchement* of the vertebrata led progressively from simpler to higher orders, genera, and species. The other three *embranchements*—the radiata, the articulata, and the molusca—did not exhibit the signs of a divine plan, and so one might even be led to think their alterations over geological epochs due to the ordinary course of causally efficient nature. But within the archetype of the vertebrates, "the constantly increasing similarity to man of the creatures that were successively called into existence, makes the final purpose obvious toward which these successions are rising."[59] The vertebrates demonstrated the willful inter-

57. Richard Owen, "On British Fossil Reptiles," *Edinburgh New Philosophical Journal* 33 (1842): 65–88. Mention of Agassiz and Vogt's find, scored by Darwin in his copy of the journal, occurs on p. 83: "The almost universal prevalence of the more or less biconcave structure of the vertebrae of the earlier reptiles, thus establishes a most interesting analogy between them and the earlier stages of growth of existing reptiles. A similar analogy has been pointed out by M. Agassiz, between the heterocercal fishes, which exclusively prevail in the oldest fossiliferous strata, and the embryos of existing homocercal fishes, which seem to pass through the heterocercal stage." Darwin's copy is held in the Manuscript Room of Cambridge University Library.

58. Louis Agassiz, "A Period in the History of our Planet," *Edinburgh New Philosophical Journal* 35 (1843): 1–27; quotation, p. 8. Darwin's copy is held in the Manuscript Room of Cambridge University Library.

59. Ibid, p. 5.

vention of the Creator in the fulfillment of his plan. Agassiz accordingly argued that extinct creatures by no means represented, as it were, mistakes for which further creations had to compensate; rather a law governing these progressive stages of replacement could be traced out in contemporary embryological recapitulations of those ancient antecedents. As he summed up his position in his 1850 book *Lake Superior* (in a passage that Darwin marked in his own presentation copy): "I can now show, through all classes of the animal kingdom, that the oldest representatives of any family agree closely with the embryonic stages of the higher types of the living representatives of the same families."[60]

Darwin became increasingly convinced that though the evidence from fossils was fragmentary, Agassiz was fundamentally right about the fact of a parallel between embryonic stages and the progressive transitions of fossilized antecedents, even if the divinely inspired naturalist were misdirected about the exact cause of the parallel (i.e., not the immediate craft of the Divinity, but for Darwin, the law of natural selection). Darwin's 1844 Essay reflects the impact on his thought of Agassiz's few fossil fish and the embryos of their descendants.

In the 1844 Essay, after giving several apparent instances of recapitulation, Darwin states that Owen and Müller object to the principle. And Darwin himself furnishes an example that appears to confirm this rejection, namely, that embryos of certain crustaceans are more complex than the adults. He thus appears to scuttle the principle of recapitulation, which seems to require adults to have progressively advanced beyond their embryos. But this is only the first move of a powerful strategy that Darwin practices throughout the 1844 Essay and the *Origin,* a strategy, however, that has misled some historians, namely, his tactic of presenting a strong set of difficul-

60. Louis Agassiz, *Lake Superior: Its Physical Character, Vegetation, and Animals* (Boston: Gould, Kendall and Lincoln, 1850), p. 262. Darwin's well-marked copy is held in the Manuscript Room of Cambridge University Library.

FIGURE 23. Louis Agassiz (1807–1873), after he had migrated to the United States; photograph taken about 1860.

ties and then demonstrating how his theory has the force to overcome them. For in the next sections of the Essay, he shows how natural selection breaches such objections as that of embryos of crustaceans being more complex than the adults. With crustaceans as an example, he demonstrates how his theory can explain certain embryological facts that have been well established.

The first fact of general agreement is the similarity of embryos of different species. Darwin proposes that if selection supervenes, not in the embryo or early in youth, but in the more mature stages, then embryos of two closely related species or breeds should resemble one another more than their adult forms would—which is in fact the case. Further, if two breeds of a species become the parent stocks of new species, then the embryos of those new species will continue to resemble each other while selection gradually separates the adult forms. Darwin recognizes what no previous recapitulationist ever denied: that though the embryos of two species might be morphologically similar, the hidden seeds of hereditary difference must yet lie within them, so as to produce differing adult types. He adds to this commonplace, however, the corollary that hereditary traits will generally appear in the offspring at those times of life when they were originally expressed in the parent. This means, of course, that morphological change will not usually be shoved far back into embryogenesis but added at the end of development. Thus these considerations show "how the embryos and young of different species might come to remain less changed than their mature parents."[61] (In the several years before the publication of the *Origin,* Darwin would conduct measurement experiments on a variety of breeds of dogs, horses, and pigeons to demonstrate that resemblances were much closer among the neonates of related breeds than among their parents.) In the case of certain crustaceans that seemed to counter the recapitulation principle, it was likely that en-

61. Darwin, "Essay of 1844," *Foundations of the Origin of Species,* p. 225.

vironmental circumstances had fostered a simpler morphology on the adults than they might have otherwise exhibited—not that the embryos were more complex, but that the adults became less complex.

Thus far, we see Darwin has preserved a set of facts that von Baer would not have denied, though of course he did so in terms of a theory of evolution that initially von Baer rejected in its German version and would continue to reject after he learned of the Englishman's new twist.[62] But Darwin went further and carved his initials on the very heart of the recapitulation hypothesis. In the last passage of the embryological section of his 1844 Essay, he considers the implications of the embryological facts he has set out, mentions other evidence, and advances a hypothesis "with much probability":

> It follows strictly from the above reasoning only that the embryos of ⟨existing⟩, for instance existing ⟨mammifers⟩ vertebrates, resembles more closely the embryo of the parent-stock of this great class, than with full-grown existing ⟨mammifers⟩ vertebrates resembles their full-grown parent-stock. But it may be argued with much probability that in the earliest & simplest condition of things that the parent & embryo would always resemble each other, & that the passage through embryonic forms is entirely due to ⟨modifications⟩ subsequent variations affecting only the ⟨later states⟩ more mature periods of life. If so, the embryos of the existing ⟨mammifers⟩ vertebrata will shadow forth the full grown structures of some of the ⟨earliest parent-stocks⟩ forms of this great class, (which existed at the earliest period of the earth's history). ⟨Such parent-stocks must have existed at necessarily the most remote epoch⟩; and accordingly animals with a fish like structure ought

62. In his later years, von Baer adopted a quasi-transmutational theory according to which certain original types (presumably of divine origin) would be transformed into a myriad of related species. See Timothy Lenoir's interesting discussion of von Baer's later ideas in *The Strategy of Life* (Chicago: University of Chicago Press, 1989), pp. 246–75.

to have preceded Birds & mammals; & of fish, that higher organised division with the vertebrae extending into a division of the tail, ought to have preceded the equal-tailed, because the ⟨young⟩ embryos of the latter have an unequal tail; & of crustacae, the entomastracae ought to have preceded the ordinary crabs barnacles—polypes ought to have preceded jellyfish, & infusorial animaculae before both. This order of precedence in time some of these cases is believed to hold good; but I think that our evidence is so exceedingly incomplete regarding the number & kinds of organisms that have existed during all, especially the earliest periods of the earth's history, that I should put no stock on this accordance, even if it held truer than it probably does in our present state of knowledge.[63]

Darwin always stayed cautious about the empirical evidence for any claim he would make; but as this passage shows, Agassiz's fish tails emboldened him to advance an argument "with much probability," namely: if adults and their embryos were similar in earlier times and if embryos have changed little over evolutionary history, then embryogenesis of present-day organisms would recapitulate earlier "full grown structures," that is, earlier *adult forms.* The archetype is the ancestor, and it is an adult.

We need remark two further aspects of Darwin's fish-tailed inspired conclusion, the face of which stares back into the nineteenth century instead of looking forward toward us. First, as the above-quoted passage indicates, Darwin is still thinking of the archetypes as independent, with each being capable of evolution and reminiscent recapitulation. But he does not yet argue that the four archetypes arise from a com-

---

63. Charles Darwin, "Essay of 1844," DAR 7, MS pp. 163b–164. I have quoted the original draft of the Essay rather than the printed version to give more of the flavor of Darwin's immediate views at the beginning of 1844, when the Essay was composed. The passage does not differ essentially from the printed version, for which the fair copy was made in September of 1844.

mon ancestor.[64] Owen's anatomical authority restrained even Darwin. In time, though, Darwin would urge the view that originally only one or a very few forms were breathed into life whence the others diverged. (Haeckel, more the traditional morphologist, would maintain that each archetype had an independent monadic source which had been spontaneously generated.)[65] Second, and remarkably, the passage demonstrates the depth of Darwin's commitment to recapitulation; that is, while the older recapitulationists offered only fragmentary parallels between embryonic states and permanent forms of lower organisms, Darwin here suggests that during development the embryo takes on virtually the full complement of the mature morphology of ancient organisms. He was a more thoroughgoing recapitulationist than his predecessors!

Owen, Chambers, and Milne-Edwards: 1844–1846

In the decade after completion of the 1844 Essay—for which he left provision in his will that it be published should his chronic illness reach its final term before his theory did— Darwin continued intermittently to work on embryology, es-

64. As late as his 1844 Essay Darwin maintained that if we considered the animal and plant archetypes together, "all the organisms *yet discovered* are descendants of probably less than ten parent-forms." See Darwin, "Essay of 1844," *Foundations of the Origin of Species,* p. 252. Darwin's belief in several distinct archetypal ancestors as the originating sources of species indicates the non-neo-Darwinian character of his theory. He reiterated his conviction in the *Origin* that "animals have descended from at most only four or five progenitors, and plants from an equal or lesser number." Analogy, he said, yet led him to think that "probably all the organic beings which have ever lived on this earth have descended from some one primordial form, into which life was first breathed." See Charles Darwin, *On the Origin of Species* (London: Murray, 1859), p. 484.

65. See Ernst Haeckel, *Generelle Morphologie der Organismen* (Berlin: Reimer, 1866), 1: 205: "each of the three kingdoms [i.e., animal, plant, and protista] consists of several stems, each of which has evolved from a particular kind of monera." The stems or phyla for the animal kingdom were the standard archetypes, though with a division of the radiata into the coelenterata and echinodermata.

pecially in relation to Owen's ideas. For example, just after he had the fair copy made of his 1844 Essay in September, he further considered another of those facts advanced by Owen that apparently refuted the recapitulation hypothesis, namely, that spiders and crabs seemed to go through virtually no transitional stages from the amorphous egg to the structured adult—according to Owen, the adult structures appeared virtually from the beginning of embryogenesis.[66] In November Darwin contrived, perhaps inspired by the just-published *Vestiges of the Natural History of Creation,* to explain this apparent failure of recapitulation. He would retain this explanation in the *Origin of Species:*

> Nov/44/ ⟨Think over this⟩
> In embryology, if embryo passes through several successive stages, then my explanation requires that the embryonic stages should have become ⟨⟨possibly⟩⟩ larger & larger ⟨⟨& more complicated⟩⟩ for by this, the selection only altering the form after birth, will allow a succession of forms to be passed through before birth. If embryonic period be shortened, eg if mammal born under fish form, then the fish form would have to be altered to conditions of life & thus reptile form might be lost if born in shorter time & then again in shorter. the mammal form is encroached on the reptile & then on the fish form, & the transition would have to be more & more rapid to be lost, *for selection might fall* on the primary vesicle & every successive transmitted stage be absolutely lost & destroyed—⟨applicable to spiders & crabs⟩[67]

Darwin seems here to have meant that over the course of species evolution, the embryo would preserve in its own ontogenetic development the forms of its ancestors, thus ever increasing the time and complexity of embryogenesis. How-

66. Owen brought this evidence against species evolution and the recapitulation theory; see my discussion above.
67. Charles Darwin, note dated November 1844; DAR 205.6, MS 34.

FIGURE 24. Sir Richard Owen, who advanced the vertebral theory of the skull; mezzotint done in 1889. Wellcome Institute Library, London.

ever, if for some reason the duration of embryogenesis were shortened, then some transitional forms would begin to fall out—for example, if a mammal were born after a gestational period that might ordinarily have only reached the embryonic stage of its fish ancestor, then the intervening reptile form would fall out. So a comparable but exaggerated foreshortening might explain some of those unusual cases—for instance, spiders and crabs—which Owen particularly cited as refuting recapitulation theory.

While Darwin thought he could handle Owen's objections to recapitulation, he began to understand how his rival's own developing theory of the archetype could be turned to evolutionary advantage. Owen agreed with Cuvier and von Baer that the animal kingdom could be united into four distinct types, that is, archetypes—the articulata, radiata, mollusca, and vertebrata. But to establish concretely that these plans governed the animal realm, researchers had to pierce through the variability of organisms to their nature, or *Bedeutung,* as Owen put it in his Germanophilic way.[68] To make this possible he sought, in his "Report on the Archetype and Homologies of the Vertebrate Skeleton" (1846), to develop a nomenclature for all the bones of the vertebrata. This required, however, that comparisons be made up and down the vertebrate line in order to give the same bones, no matter in what organisms they appeared, the same names. Thus, he proposed that the bones in the pectoral fin of fish, in the paddle of the porpoise, and in the arm of man be given the same names since they were homologously similar—that is, had the same "nature"—but only analogous to structures in the

68. Richard Owen, *On the Nature of Limbs* (London: Van Voorst, 1849), p. 2: "I have used . . . the word 'Nature' in the sense of the German 'Bedeutung,' as signifying that essential character of a part which belongs to it in its relation to a predetermined pattern, answering to the 'idea' of the Archetypal World in the Platonic cosmogony, which archetype or primal matter is the basis supporting all the modifications of such part for specific powers and actions in all animals possessing it, and to which archetypal form we come in the course of our comparison of those modifications, finally to deduce their subject."

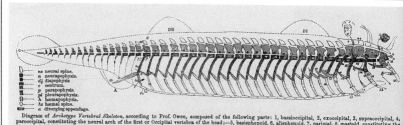

Diagram of *Archetype Vertebral Skeleton*, according to Prof. Owen, composed of the following parts: 1, basioccipital, 2, exoccipital, 3, supraoccipital, 4, paroccipital, constituting the neural arch of the first or Occipital vertebra of the head;—5, basisphenoid, 6, alisphenoid, 7, parietal, 8, mastoid, constituting the neural arch of the second or Parietal vertebra;—9, presphenoid, 10, orbitosphenoid, 11, frontal, 12, postfrontal, constituting the neural arch of the third or Frontal vertebra;—13, vomer, 14, prefrontals, 15, nasal, constituting the neural arch of the fourth or Nasal vertebra;—16, petrosal, or acoustic sense-capsule;—17, sclerotic, or ophthalmic sense-capsule; 18, 19, ethmoturbinal, or olfactive sense-capsule;—20, palatine, 21, maxillary, 22, premaxillary, 23, 24, 25, pterygoid, 26, malar, 27, squamous, constituting the hæmal arch of the Nasal vertebra with its diverging appendages;—28, tympanic, 29–32, mandibular or lower jaw, 34–37, operculars, constituting the hæmal arch of the Frontal vertebra, with its diverging appendages;—38, stylohyal, 39, epihyal, 40, ceratohyal, 41, basihyal, 42, glossohyal, 43, urohyal, 44, branchiostegal, constituting the hæmal arch of the Parietal vertebra, with its diverging appendages;—50, suprascapula, 51, scapula, 52, coracoid, 52′ episternum, 53, anterior member, constituting the hæmal arch of the Occipital vertebra, with its diverging appendages;—58, clavicle, the hæmapophysis of the Post-occipital vertebra;—62, ilium, 63, ischium, 65–69, posterior extremity, constituting the hæmal arch of the Sacral vertebra S, with its diverging appendages; 64, pubis, the hæmapophysis of one of the lumbar (?) vertebræ. The chief osseous developments of the dermo-skeleton are added in outline, as the median horn supported by the nasal spine, 15, in the Rhinoceros; the pair of lateral horns developed from the frontal spine, 11, in most Ruminants; the median dorsal fins, D I, D II, in Fishes and Cetaceans; the caudal fin, c, and the anal fin, A, in Fishes.

Figure 25. Illustration of Owen's construction of the vertebrate archetype, from William Carpenter, *Principles of Comparative Physiology* (1854).

"arm" of the crab. On the basis of such homologous comparisons—whose ultimate outcome had been ordained by his transcendental persuasion—Owen reconstructed the ideal plan of the vertebrata (see fig. 25). The plan was simply that of a backbone, a vertebral column whose different processes were modifiable into tail, ribs, pelvis, limbs, and head.[69] In the "Report" Owen endorsed Oken's notion that the vertebrate skull was merely a continuation and modification of the vertebrae. In formulating his argument for Oken's theory,

69. Owen derived his theory of the vertebrate archetype from several sources. Cuvier, of course, maintained that the animal kingdom exhibited four exclusive body types, the radiata, articulata, mollusca, and vertebrata. Owen's mentor Green expressed the nut of the theory when he credited (*Vital Dynamics*, p. 57) Oken and Carus with developing Goethe's insight that the variegated parts of plants and animals were only modifications of a few fundamental forms (see chap 3, above), that, "for instance, the osseous system in every part, and in its most complicated total result, is but the repetition of a simple vertebra." The most proximate source for Owen's theory of the archetype, though, must be von Baer's *Entwickelungsgeschichte der Thiere* and Barry's articles on von Baer's theory (see n. 20, above).

he mentioned that the human fetal skull displayed distinct pieces of not-yet ossified bones. "These and the like correspondences," he observed, "between the points of ossification of the human foetal skeleton, and the separate bones of the adult skeletons of inferior animals, are pregnant with interest, and rank among the most striking illustrations of unity of plan in the vertebrate organization."[70] Both his endorsement of Oken's theory and his cautiously faint appeal to a recapitulation *within* an archetype would later give Huxley occasion to exercise his passionate disdain for the Hunterian Professor.

In the meantime, though, Darwin turned Owen's work to evolutionary advantage by insisting (in his penciled reading notes to the "Report") that the vertebrate archetype must have actually existed as an ancestor: "For myself," Darwin wrote, "I cannot any more doubt that some primordial race existed long anterior to the lowest part of the Silurian system, which had a chain of vertebrae and no skull, than I doubt the correctness of the nature of the skull & the beautiful explanation it affords of so many bones in it, even in adult Birds' skull."[71] For Darwin, Owen's vertebrate archetype was thus not merely an ideal plan that conceptually united all vertebrate species; it was an actual creature that reproductively united its descendant species in real genealogical relations.

Darwin's transformation of Owen's archetype into a living creature combined two different traditions in comparative anatomy, that of the functionalists, like Cuvier, and that of the ideal morphologists, like Owen. Cuvier had explained the internal purposiveness of his four *embranchements* as a function of the similar modes of life exercised by species members within the four organizational types. Ultimately the correlation of parts that each of the types expressed resulted

---

70. Richard Owen, "Report on the Archetype and Homologies of the Vertebrate Skeleton," *Report of the Sixteenth Meeting of the British Association for the Advancement of Science; Held at Southampton in September 1846* (London: Murray, 1847), pp. 169–340; quotation from p. 272.

71. Charles Darwin, DAR 74, MS 113.

from adaptations to their respective conditions of existence. Owen, developing the Goethe-Carus-Green conception of organizational types, focused on internal purposiveness. He acknowledged that transformations of homologous parts within an archetype had functional significance in relation to the external environment, but such adaptations could not explain the fundamental traits of the type itself. For instance, the particular features of the human limb, say, the disposition of the fingers, undoubtedly served functional ends; but the homologous pattern of bones displayed in the fin of the whale could hardly be given the same kind of functional account. The type of the limb itself, the "general homology" in Owen's terms, remained neutral to the particular functions of the human limb and the whale fin.[72] The archetypal pattern, in Owen's view, thus could not be explained according to the functionalism of Cuvier, nor according to the teleological mode of Paley's natural theology, for that matter. Darwin would agree with Owen that the type itself could not be conceived as an immediate adaptation to the various environments in which vertebrates would be presently found. Darwin's genius was to see that the vertebrate pattern could yet be explained as the inheritance from common ancestors, living in the long past, that did acquire the pattern as a series of functional adaptations. In making the archetype the ancestor, Darwin historicized and thereby naturalized teleology.

Almost immediately after Darwin had completed his 1844 Essay, with a fair copy made in September of that year, Robert Chambers's anonymously published *Vestiges of the Natural History of Creation* landed with a large splash in the midst of the scientific community. Darwin quickly purchased a copy and began reading it in November. The book argued for species evolution in a way that made him uncomfortable, chiefly because of its similarity to his own considerations. In the central argument, Chambers elaborated the idea of a unity of

72. See Owen, "Report on the Archetype," p. 241, and *On the Nature of Limbs,* pp. 39–40.

plan running through the animal kingdom. That plan bespoke development. Embryology revealed this most perspicuously, for "physiologists have observed that each animal passes, in the course of its germinal history, through a series of changes resembling the permanent forms of the various orders of animals inferior to it in the scale."[73] The "fact" of recapitulation then became for Chambers a principle by which species development could be explained. In the past, if a lower species, say, a kind of fish, were to "protract . . . gestation over a small period," the next-highest type would be reached in embryonic development and then would be born a higher species. Thus, as Tiedemann and Meckel (whom Chambers cited) had argued, the same law that governed the development of the embryo also governed the development of species. This conception was close enough to the one Darwin had worked out in his "Notebook E" that it must have made him more than a little concerned, particularly when the crashingly negative reaction of critics flowed back over the *Vestiges* and its evolutionary doctrine. After all, in Darwin's view recapitulation preserved past progressive stages and served as the platform for future advance: the last stage in the sequence, as it were, appearing *in potentia* as the embryo springs from the womb to face the selective pressures of a new or changing environment. And both Darwin and Chambers pictured this progressive advance as

73. [Robert Chambers], *Vestiges of the Natural History of Creation,* 3d ed. (New York: Harper Brothers, n.d.), pp. 102–3. In the first edition of 1844, Chambers simply endorsed the full-blown recapitulationist thesis that the embryo passed through the "adult" stages of lower organisms. After he read Carpenter's account of von Baer's law, however, he added a passage that stated the embryo went through stages comparable to the embryonic (not adult) stages of organisms below it (p. 110). Chambers did not bother to reconcile different parts of his text in subsequent editions; and so in his book, and perhaps also in his head, the recapitulationist and von Baerean doctrines rested side by side. James Secord describes the uneasy development of Chambers's theories in "Behind the Veil: Robert Chambers and *Vestiges,*" in *History, Humanity, and Evolution,* ed. James Moore (Cambridge: Cambridge University Press, 1989), pp. 165–94.

moving up through a ramified tree structure—to quote Carpenter's representation of Chamber's system: "from the germs first introduced . . . the whole succession of organic forms, from the simplest Protophyte up to the Oak or Palm, from the Protozoon up to Man, has been gradually evolved; not, however, in a single series, but from several distinct *stirpes* [i.e., lines or phyla] whose development has taken different directions."[74] Carpenter and the host of British naturalists utterly rejected this evolutionary thesis. Darwin had cause to be worried. He even became wary of the use of the term "development" for his theory, so powerful a symbol of Chamber's maligned hypothesis had the word become.[75]

When in the next year Chambers produced a sequel that replied to critics, Darwin took notes on the added evidence for the principle of recapitulation. He seems to have been particularly interested in the cast of naturalists—Owen, Hunter, and Roget especially—whom Chambers recruited in defense of the principle. But the work that made the deepest impression on Darwin at this time was Henri Milne-Edward's essay on natural classification, which he read in December of 1846 and regarded as "the most profound paper I have ever seen on affinities."[76]

In his "Considérations sur quelques principes relatifs à la classification naturelle des animaux" (1844), Milne-Edwards (1800–1885) suggested the utility of embryological study for determining the natural affinities of animals. He argued that the stages through which the embryo passed revealed zoological relations that were often disguised in the adult organism. His thesis, as he admitted, was virtually the same as von

74. William Carpenter, *Principles of Comparative Physiology,* 4th ed., p. 106. See also Chambers, *Natural History of Creation,* p. 223.

75. See Charles Darwin to Leonard Jenyns (14 February 1845), *Correspondence of Charles Darwin,* 3: 143. Jon Hodge called this letter to my attention.

76. Darwin's reading notes on Chambers are in DAR 205.9 (2), MS 215, and on Milne-Edwards in DAR 72, MS 117–22. Ospovat discusses the impact of Milne-Edwards in his *Development of Darwin's Theory,* pp. 159–65, but arrives at conclusions rather different from mine.

Baer's, though independently arrived at, namely: that the "modifications which are successively displayed in the constitution of the young animal or of the germ from which they will emerge are those that successively determine its existence as a member of its *embranchement,* its class, its family."[77] Unlike the recapitulationists (for whom, however, he had great respect), Milne-Edwards did not believe that the embryo moved through the "species" forms of inferior creatures, but that it did rise from the more general to the more particular forms of a given species. This meant that it was not necessary to reject, as his own teacher Cuvier did, "the entire idea of a natural classification corresponding to the different degrees of perfection of animated beings." Rather one only had to reject the notion that it was possible "to represent zoological affinities with the help of one diagrammatic line."[78] The affinities should rather be represented as branches of a tree, and embryos of, say, the same genus of mammal would begin, as it were, rooted in a morphologically similar germ, climb together a trunk from which would branch the other major types, and crawl on to limbs representing the same family, class, and genus, until finally they reached the species fork.[79] Embryonic recapitulation in this sense (that is, from homogeneous to heterogeneous within a particular archetype) conformed, not to a linear series of perfection—against which virtually all leading zoologists contended—but to a branching series.

Darwin was antecedently convinced of this much even before he read Milne-Edwards, though the eminent Belgian zoologist confirmed these ideas for him. Milne-Edwards, however, added a consideration that, while again similar to one made by von Baer, yet struck Darwin full strength. This was the way in which "higher" and "more perfect" species

77. Henri Milne-Edwards, "Considérations sur quelques principes relatifs à la classification naturelle des animaux," *Annales des sciences naturelles,* 3d ser. 1 (1844): 65–99; quotation from p. 68.

78. Ibid., pp. 70–71.

79. Ibid., p. 72.

could be measured. Milne-Edwards rejected the idea, for example, that those radiata were more perfect as they approached binary animals. Rather, within the radiata, higher levels of perfection were reached by those species that more completely adapted their basic structures to the economy of nature through "the division of physiological labor."[80] For Darwin this approach provided a standard (though not the only one) by which the perfection of animals in their evolutionary trajectory could be measured, as he explained to his friend Joseph Hooker sometime later:

> within the same kingdom I am inclined to think that "highest" usually means that form which has undergone more "morphological differentiation" from the common embryo or archetype of the class . . . The specialization of parts to different functions, or "the division of physiological labour" of Milne Edwards . . . is the best definition.[81]

Though recently several historians and philosophers of biology have argued that the branching conception of evolution precludes notions of progression, it is quite clear from the above quotation that this was not Darwin's understanding. On the contrary, for Darwin evolution meant the advancement of higher, more perfect types—not, however, in a single linear train, but along different branches of the evolutionary tree, each arching up from a common stem to reach higher levels of adaptive organic differentiation. Darwin discovered, however, that his way of scaling perfection was not without difficulty. His exasperating affair with barnacles proved that.

The Embryology of Barnacles and the Criteria of Progressive Development: 1846–1854

In October 1846 Darwin wrote his friend Joseph Hooker of an investigation he had decided to undertake, one that would usher him onto his favorite ground:

---

80. Ibid., p. 76.
81. Charles Darwin to Joseph Hooker (27 June 1854), *More Letters of Charles Darwin,* ed. Francis Darwin (London: Murray, 1903), 1: 76.

> I am going to begin some papers on the lower marine animals, which will last me some months, perhaps a year, & then I shall begin looking over my ten-year-long accumulation of notes on species & varieties which, with writing, I daresay will take me five years, & then when published, I daresay I shall stand infinitely low in the opinion of all sound naturalists—so this is my prospect for the future.[82]

Initially an unusual species of barnacle, which he had discovered off southern Chile, piqued his evolutionary interest. The tiny *arthrobalanus* would drill through the shell of a passing mollusk and dwell happily therein instead of secreting its own shell and attaching itself to a rock. Darwin's project thus began modestly, but by the time he had finished some eight years later, he had described all the known living and fossil cirripides. With the last of his four volumes on barnacles published in 1854, he had worked up a great hatred for these small, complicated, and slimy creatures.[83] The study, however, helped confirm him in his belief that embryological development displayed phylogenetic history.[84]

In each of the two families of living cirripidae, Lepadidae and Balanus, the free-swimming larvae go through two stages.[85] From the egg and prior to attachment, these animals develop a long narrow anterior part, which displays two prehensile antennae and two simple eyes, and a rounder, blunt

82. Charles Darwin to Joseph Hooker (2 October 1846), *Correspondence of Charles Darwin,* 3: 346.

83. Charles Darwin, *A Monograph of the Fossil Lepadidae or, Pedunculated Cirripedes of Great Britain* (London: Ray Society, 1851); *A Monograph of the Sub-Class Cirripedia, with Figures of all the Species: The Lepadidae or, Pedunculated Cirripedes* (London: Ray Society, 1851); *A Monograph of the Fossil Balanidae and Verrucidae of Great Britain* (London: Palaeontographical Society, 1854); *A Monograph of the Sub-Class Cirripedia: The Balanidae (or Sessile Cirripedes), the Verrucidae, &c* (London: Ray Society, 1854).

84. The very best account of Darwin's barnacle work is by Marsha Richmond. See her "Darwin's Study of Cirripedia," in appendix 2 of *Correspondence of Charles Darwin,* 4: 388–409.

85. These larval stages are described in Darwin's *Monograph of the Sub-Class Cirripedia: The Lepadidae,* pp. 8–67.

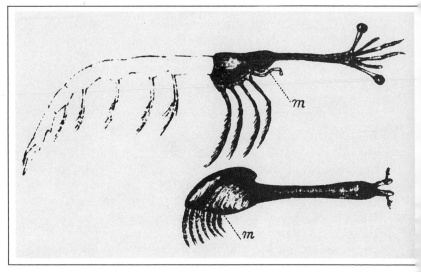

FIGURE 26. Illustration of the homologies between a Stomapod crustacean (with posterior on sketched) and a Lepas barnacle larva (*m* indicates a mouth), from Charles Darwin, *A Monograph the Sub-Class Cirripedia: The Lepadidae* (1851).

posterior, from which extend three pairs of legs and a mouth just in front of the first pair (see fig. 26). At the end of the second stage of their development, the larvae secret a cement through their antennae and attach themselves to some object. Darwin, much like Swammerdam before him, could dissect out the mature animal within the integuments of the recently attached larva. In due course the "skin" would calcify, and the living adult would dwell with its cephalic end down against the attachment, while its posterior would wave in the breeze

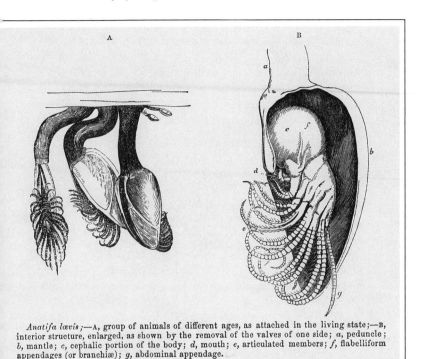

Anatifa lævis;—A, group of animals of different ages, as attached in the living state;—B, interior structure, enlarged, as shown by the removal of the valves of one side; *a*, peduncle; *b*, mantle; *c*, cephalic portion of the body; *d*, mouth; *e*, articulated members; *f*, flabelliform appendages (or branchiæ); *g*, abdominal appendage.

FIGURE 27. Illustrations, drawn from Darwin, of adult barnacle of the order Lepadidae, from William Carpenter, *Principles of Comparative Physiology* (1854).

of the water flowing by the shell opening (see fig. 27). The adult became more simplified in its structures and lost its distinctive signs of zoological affinity. Indeed, Georges Cuvier had classified cirripidae within the mollusk type because of the character of their adult stage. But in the 1830s, when the larvae had been identified, they appeared to be more properly located in the articulata, as a subclass of crustaceans, where Darwin placed them.

Darwin's own study of larval development convinced him

of the great value of embryogenesis for properly understanding the systematic relationships exhibited by different classificatory types. The larvae of barnacles reflected, as he carefully illustrated in his monograph (see fig. 26), the same structures as *adult* crustaceans. In the *Origin of Species* he would insist on this value of embryology for providing a guide to phylogenetic descent relationships:[86]

> Descent being on my view the hidden bond of connexion which naturalists have been seeking under the term of the natural system. On this view we can understand how it is that, in the eyes of most naturalists, the structure of the embryo is even more important for classification than that of the adult. For the embryo is the animal in its less modified state; and in so far it reveals the structure of its progenitor.[87]

This passage from the *Origin* clearly indicates the importance for Darwin of recapitulation for systematic classification and its status as the causal outcome of species descent. Darwin, though, would not disclose his full-blown species heresy in the barnacle books, even if he unobtrusively used evolutionary recapitulation as the principle to place barnacles within the subclass of crustaceans. But in the barnacle books Darwin also tipped his hand on another significant feature of his secret theory—that of the progressive status of these animals.

The Cuvierian doctrine of types as glossed by von Baer and Owen did allow one to speak of higher and lower species *within* the archetype (i.e., greater or lesser morphological differentiation from the archetype). However, that doctrine specifically forbade placing the archetypes on a progressive scale—since that bespoke the pernicious ideas of "unity of organization" and ultimately of "transmutation."[88] Darwin, as

86. Charles Darwin, *On the Origin of Species* (London: Murray, 1859), pp. 440–41, 449.

87. Ibid.,p. 449.

88. See the discussion above of Owen's linking of the doctrines of "unity of organization" and "transmutation."

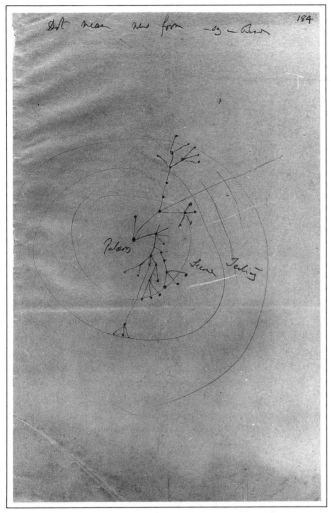

FIGURE 28. Sketch of von Baer's theory of development within an archetype, from Darwin's manuscript notes, ca. 1853–1854.

we have seen, embraced Milne-Edward's (which in his barnacle volume of 1854 he identified as von Baer's)[89] comparable criterion for intratype progression (i.e., greater or lesser morphological differentiation through embryological stages; see Darwin's effort to work out von Baer's criterion in fig. 28). Since cirripides had, however, undergone considerable differentiation in the adult—yet of a simplifying sort— he appended a more general criterion of progression:

> On the whole, I look at a cirripede as being of a low type, which has undergone much morphological differentiation, and which has, in some few lines of structure, arrived at considerable perfection— meaning, by the terms perfection and lowness, some vague resemblance to animals universally considered of a higher rank.[90]

Darwin needed the extra note of comparison with animals "universally considered of a higher rank," since the criterion of embryological divergence alone would have required him to rank the adult barnacle as "higher," even though it was less complex than its own larva. He had actually worked out this dual criterion of progress—that is, embryological divergence plus comparison with higher animals—at an earlier time. In a loose note dated March 1845, he reflected:

> What is the highest form of any class? Not that which has undergone most changes. for changes may reduce organization:—generally, however, that which has undergone most changes & which approaches nearest to man (?) Hardly applicable to insects or plants—Each perfect for its end, & not most perfect— "complication not of homologous organs". "combined, when comparison with man as a model" Perfect adaptation comes in idea[91]

89. Darwin, *A Monograph of the Sub-Class Cirripedia: The Balanidae,* p. 20.
90. Ibid.
91. Darwin, loose note, DAR 205.9 (2), MS 200.

Early on he thus recognized—perhaps prompted by Agassiz's idea that progressive development was to be found only in the vertebrates—that, at least when comparisons were within an archetype (and thus "homologous organs" were at issue), reference had ideally to be made to "higher organisms," and, in the case of the vertebrates, most usually to the highest organism, man.

With his dual criterion for progress, Darwin would be able, later in the *Origin of Species,* to explain why the simpler character of the adults of some species, like those of barnacles, did not refute the general principle of recapitulation, as von Baer and Owen had argued it did. For natural selection might have backfilled the gap in the economy of nature, and "in this case the final metamorphosis would be said to be retrograde."[92] Quite obviously Darwin regarded his evolutionary theory as a guarantee of general progress, even if in some notable instances one had to admit a slide to the more primitive.

Huxley's Objections to Recapitulation and Darwin's Experiments

Almost immediately after the final volume of the barnacle opus had been published, Darwin set out to experiment in embryology. In spring of 1855, he sought the help of his cousin William Fox, a man of good nature who could nonetheless "kill babies."[93] Darwin wanted to measure the skeletons of newly hatched pigeons of different breeds. Soon he himself was doing the "black deed," as he wrote Fox in July, and had just "murdered an angelic little fantail and pouter at 10 days old," gassing them with prussic acid.[94] His research extended beyond pigeons, as he further related: "I have puppies of Bull-dogs & Greyhound in salt.—& I have had

92. Darwin, *Origin of Species,* p. 448.

93. Charles Darwin to William Fox (27 March 1855), *Correspondence of Charles Darwin,* 5: 294.

94. Charles Darwin to William Fox (22 July 1855), *Correspondence of Charles Darwin,* 5: 386.

Carthorse & Race Horse young colts carefully measured."[95] All this slaughter, he told his cousin, was "to ascertain, whether the young of our domestic breeds differ as much from each other as do their parents."[96]

A series of disagreements with a new and highly respected friend undoubtedly spurred Darwin to demonstrate experimentally that of which he was antecedently convinced. In April of 1853, two years after their correspondence had begun, Darwin wrote Thomas Henry Huxley that he felt distressed over some remarks his friend had made in a paper on invertebrate morphology. In that article, Huxley undertook extensive anatomical investigation in order to distill out the archetype of the Cephalous Mollusca. He maintained that the Cephalopoda and Gasteropoda were morphologically one, but that no transitional forms linked them to other mollusks, such as the ascidians. He thus concluded that there was "no progression from a lower to a higher type, but merely a more or less complete evolution of one type."[97] Darwin responded to the paper by gently suggesting that perhaps the differences Huxley had found among mollusk types might actually be no greater than variations within one species (and that therefore transitions would be actually quite easy). He confessed that he had believed "the archetype in imagination was always in some degree embryonic, and therefore capable [of] and generally undergoing further development."[98] Darwin's brief

95. Charles Darwin to William Fox (7 May 1855), *Correspondence of Charles Darwin,* 5: 326. In the Darwin papers there exist many sheets of Darwin's calculations on embryos and parents. For example, DAR 205.6, MS 10, is a table of measurements comparing foals and dams of cart horses and brood horses.

96. Charles Darwin to William Fox (23 May 1855), *Correspondence of Charles Darwin,* 5: 337.

97. Thomas Henry Huxley, "On the Morphology of the Cephalous Mollusca" (1853), in *The Scientific Memoirs of Thomas Henry Huxley,* ed. Michael Foster and E. Ray Lankester (London: Macmillan, 1898), 1: 152–93; quotation from p. 192.

98. Charles Darwin to Thomas Henry Huxley (23 April 1853), *Correspondence of Charles Darwin,* 5: 133–34.

response reveals his still potent embryological model of species evolution, in which embryogenesis pictured phylogenetic transformations: that is, the real ancestor "was in some degree embryonic" and capable of further evolution.

Huxley would not have been receptive to Darwin's suggestions at this time. He had been working on his translation of sections of von Baer's *Entwickelungsgeschichte der Thiere,* which appeared that same year, 1853; and he had been deeply impressed by its theoretical considerations, which of course opposed evolutionary recapitulation and species transformation. And the division between Darwin and his new friend would have deepened as the result of Huxley's celebrated brain-dashing review of the tenth edition of that "still notorious work of fiction," the *Vestiges of the Natural History of Creation.*[99] Huxley's review (1854) reverberated loudly through the scientific community because he not only rattled the skull of the author, whom he and Darwin rightly suspected to be Robert Chambers, but because he also bashed the authority of Agassiz and Owen.[100]

Huxley focused on two principal objections to the Vestigiarian's theory that prescribed "a gradual evolution of high from low . . . [under] divine ordination."[101] He first showed that the author muddled the idea of natural law with that of divine ordinance, so that his explanation of natural phenomena ultimately did not differ from that of Moses. Huxley heaped up the larger weight of his invectives, however, on the supposed evidence Chambers cited, particularly Agassiz's claim that the morphology of fossil fish had been recapitu-

99. Thomas Henry Huxley, "Vestiges of the Natural History of Creation, Tenth Edition" (1854), in the supplement to *The Scientific Memoirs of Thomas Henry Huxley,* p. 1–19.

100. The history of Huxley's morphological theory before and after the *Origin of Species* has been skillfully worked out for the first time by Sherrie Lyons, in her "Evolution of Thomas Henry Huxley's Evolutionary Views" (Ph.D. diss., University of Chicago, 1990).

101. [Robert Chambers], *Vestiges of the Natural History of Creation,* 10th ed. (London: Churchill, 1853), pp. vii–viii.

lated in contemporary fish embryos. Chambers had happily used Agassiz's theory of fish tails as the principal evidence for law-governed progressivism. To add exactly the right spice to this argument, which considerably sharpened Huxley's appetite, Chambers cited an anonymous article in the *Quarterly* that also appealed to Agassiz's work and advanced a similarly progressivist theory. Chambers ascribed the article to Richard Owen. And Huxley knew Chambers was entirely correct, though he maliciously feigned doubt.[102] Huxley had a feast. The bouillabaisse included the Vestigiarian, whose charlatanism sucked ten editions out of a gullible public, Louis Agassiz, whose religious fervor suffocated exact investigation, and Richard Owen, whom he was growing to detest.[103] Huxley attempted to demolish Agassiz's argument by picking out morphological characters of ancient fish that seemed advanced and piling them with those of recent fish that seemed primitive. In the end, not even the husk of a progressive theory remained. And through his analysis, Huxley chided the Vestigiarian for ascribing the supporting views of the *Quarterly* writer to Owen—for the esteemed Hunterian Professor would never utter such absurd things! Huxley cleaned the shell of the progressivist proposition with great relish.

But it was Darwin who had dyspepsia. He wrote Huxley in September of 1854, just after the review appeared, to confess that his own ideas were "almost as unorthodox about species as the *Vestiges* itself."[104] He went on to say he was "rather sorry

102. [Richard Owen], "Lyell—on Life and its Successive Development," *Quarterly Review* 89 (1851): 412–51. Huxley and others knew immediately that the article was by Owen. See Thomas Henry Huxley to W. Macleay (9 November 1851), *Life and Letters of Thomas H. Huxley,* ed. Leonard Huxley (New York: Appleton, 1900), 1: 101.

103. Huxley broke with Owen in 1857 over the established scientist's momentary usurpation of the title of professor of paleontology of the School of Mines, a move which severely encroached on Huxley's territory. Although Owen helped Huxley professionally in the early 1850s, even in 1851 the young scientist confessed that with his patron "I feel it necessary to be always on my guard." See ibid., p. 102.

104. Charles Darwin to Thomas Henry Huxley (2 September 1854), *Correspondence of Charles Darwin,* 5: 213.

FIGURE 29. Thomas Henry Huxley, 1825–1895, who resisted Darwin's theory of embryological recapitulation; portrait done in 1883.

you do not think more of Agassiz's embryological stages, for
though I saw how excessively weak the evidence was, I was
led to hope in its truth."[105]

Darwin must have felt even more queasy when Huxley, in
January of 1855, again slashed away at progressive evolution
and Agassiz's evidence. The occasion was a review of the
fourth edition (1854) of William Carpenter's *Comparative
Physiology*.[106] While Huxley generously praised Carpenter's
accomplishment, he had "a crow to pick" with him on a
foul idea that he must have found suspiciously close to
the Vestigiarian's—perhaps for good reason.[107] Carpenter,
though friendly to the *Vestiges,* rejected the hypothesis of an
evolutionary transmutation of forms but not the idea that
there might be a progressive replacement of forms according
to a divine plan. Both paleontological and embryological
evidence, particularly that supplied by Agassiz's fish tails, suggested that the earliest forms were of "archetypal generality"
and very similar to the embryological condition of those living animals to which they bore affinity.[108] Carpenter thus
concluded that this parallel development suggested a "plan
on which the progressive evolution of the great scheme of
Organic Creation has proceeded."[109] Huxley detested the
idea of progress in nature, precisely because it almost always
led to assumptions about divine meddling in the natural
world.[110] He vigorously dissented, then, from Carpenter's

105. Ibid.

106. [Thomas Henry Huxley], "Science," *Westminster Review,* 63 (n.s., 7)
(1855): 239–53.

107. While Carpenter rejected the letter of Chambers's evolutionary theory, he nonetheless, at the request of the printer (Churchill) and the anonymous author, helped straighten out the physiology and zoology of the
Vestiges for the tenth edition. Finally, Carpenter's own ideas began slipping
toward those of the Vestigiarian. See Milton Millhauser, *Just Before Darwin:
Robert Chambers and Vestiges* (Middletown, Conn.: Wesleyan University
Press, 1959), p. 147.

108. Carpenter, *Comparative Physiology,* 4th ed., p. 134.

109. Ibid., p. 143.

110. It was this difficulty with the concept of gradual progress that led Huxley to hesitate for a long time over Darwin's theory. See Lyons, "The Evolution
of Thomas Henry Huxley's Evolutionary Views."

proposal of a "great scheme" of evolution and set out to incinerate its evidence, particularly that drawn from Agassiz. He maintained that Agassiz and Vogt's embryological evidence was confined to salmon alone and that other species had homocercal tails throughout their embryological development. And even in the case of salmon, they actually had homocercal tails at the very beginning of embryogenesis, changing to heterocercality only subsequently. Thus, in Huxley's estimation, more exacting experimental evidence told against Carpenter's particular hypothesis and other efforts to support the "progressivist doctrine."[111]

Darwin obviously read Huxley's review with some anxiety, since his own progressivist doctrine enjoyed considerable common ground with Carpenter's, differing, though, in substituting natural selection for divine guidance.[112] Indeed, the similarity of their ideas made the way easy for Carpenter to adopt a modification of Darwin's evolutionary theory as soon as he read the *Origin*—and Darwin thought the eminent zoologist would come around completely when he further reflected on "Homology and Embryology."[113] But Huxley, who like Carpenter initially accepted a version of evolutionary theory after reading the *Origin,* yet still demurred precisely on the subjects of homology and embryology. Darwin thought he and his young friend were "in as fine a frame of mind to discuss the point as any two religionists."[114] Though Darwin would not retreat in the face of his friend's antievolutionary broadsides, he remained cautious about vulnerable elements of his general theory, especially recapitulation.

Darwin finally began, on 14 May 1856, to write out his gen-

111. Huxley, "Science," pp. 242–47.

112. Darwin's reading notes on Huxley's paper mostly sketch his friend's objections to Agassiz. See DAR 205.6, MS 58.

113. Charles Darwin to Charles Lyell (5 December 1859), *Life and Letters of Charles Darwin,* ed. Francis Darwin (New York: D. Appleton, 1891), 2: 35: "He [Carpenter] is a convert, but does not go quite so far as I . . . He can hardly admit all vertebrates from one parent. He will surely come to this from Homology and Embryology."

114. Charles Darwin to Joseph Hooker (14 December 1859), *Life and Letters,* 2: 39.

eral theory. He hoped to publish his growing manuscript
under the title of *Natural Selection*. As mentioned above, Darwin's work was interrupted by the reception of Alfred Russel
Wallace's letter in spring of 1858, in which this slight acquaintance revealed a theory so very close to evolution by natural
selection that Darwin plunged into despair over his apparently crushed originality. But with honor-preserving encouragement from his friends Lyell and Hooker, Darwin began
"abstracting" his by-then engorged manuscript. After reducing its considerable bulk, he added the yet uncompleted
chapters and published the *Origin of Species* the following
fall, in 1859.

While laboring on the early parts of his big manuscript,
Darwin received a letter from an American friend, the Yale
geologist and zoologist James Dwight Dana (1813–95),
whose earlier work on crustaceans Darwin had thoroughly
studied.[115] Dana wrote Darwin in September 1856 to describe a report that Louis Agassiz, who had since moved to
Harvard, had delivered to a scientific meeting. The report
supplied further evidence for the principle of recapitulation.
Agassiz displayed to the meeting some fry of the North American garpike that, as Dana described to Darwin, "had the tail of
the Ancient Ganoids—That is, the vertebrae were actually
continued to the extremity of the upper lobe—This upper
lobe . . . drops off as the animal grows & the fish then is of the
modern type of form."[116] Shortly after receiving this letter,
Darwin wrote back to say: "What a striking case of vertebrae in
tail of young Gar-Pike; I wish with all my heart that Agassiz
would publish in detail on his theory of parallelism of
geological & embryological development; I *wish* to believe,
but have not seen nearly enough as yet to make me a
disciple."[117] Having thus expressed a continued caution—

115. James Dwight Dana to Charles Darwin (8 September 1856), *Correspondence of Charles Darwin,* 6: 215–16.

116. Ibid., p. 216.

117. Charles Darwin to James Dwight Dana (29 September 1856), *Correspondence of Charles Darwin,* 6: 235.

undoubtedly enforced by Huxley—he then revealed to Dana the general features of his new theory of species change. He thought his friend would "give me credit for not having come to so heterodox a conclusion, without much deliberation."[118] Dana immediately responded with perhaps unexpected encouragement. He urged that Agassiz's principle could be subsumed under a larger progressive law which made it natural that "the history of an individual in its particulars should sometimes run parallel with that of the palaeontological history of the tribe to which it pertains." This progressive law, he proposed, "involves the expression of a type-idea in forms or groups of increasing diversity and generally of higher elevation; always resulting in a purer & fuller exhibition of the type." Darwin would undoubtedly have given this a reading congenial to his own theory, especially when Dana closed the letter with the remarkable sentiment: "I believe there is real truth in the results of your labors, and the best of foundations for general laws or principles."[119]

Dana's confidence and his intimation of comparable belief[120] must have strengthened Darwin's defenses against Huxley's relentless onslaught against species evolution and recapitulation. Darwin at least made a memorandum of Dana's views on the back of the letter: "This note contains fact of Cavern Rat being American Form & Dana's belief that in Embryonic changes & geological expression there is a certain parallelism [*sic*] from the unfolding of the type idea to its full display."[121]

Darwin would not simply rest secure with support like Dana's. He had already taken up arms against Huxley. Just a few months after Huxley's review of Carpenter appeared, Darwin set out, as I have mentioned above, to experiment in

118. Ibid., p. 236.

119. James Dwight Dana to Charles Darwin (8 December 1856), *Correspondence of Charles Darwin,* 6: 299–300.

120. Dana later openly declared for transformism—save in the case of man.

121. This annotation is recorded by the editors of *Correspondence of Charles Darwin,* 6: 300.

embryology. His aim, I believe, was to provide as much evidence for what had become, as he later explained to Joseph Hooker, "his pet bit," namely, his embryological theories.[122] He wished to show through experiment that embryos, or at least neonates, in several different species—for example, dogs, horses, and pigeons—exhibited greater "archetypal generality" than full-grown adults of related breeds. This pattern of development would suggest that for the most part, species-altering adaptations had been acquired by adult organisms and subsequently would appear only in the adults—tacked on, as it were, to the final stages of development. This would mean, therefore, that the earlier embryonic stages would represent the ancient, ancestral (adult) progenitors of current species. These experiments and the arguments for recapitulation that Huxley failed to shake loose from Darwin were compressed into chapter 13 of the *Origin of Species* and spread out in the *Descent of Man*.

Embryological Recapitulation in the Origin of Species

Those historians of science who have recently considered Darwin's embryological theories—for instance, Gould, Ospovat, Mayr, and Bowler—have maintained that Darwin rejected the idea that the embryo recapitulated the adult stages of its evolutionary progenitors. The common view holds that Darwin simply and wisely adopted von Baer's position that the embryo advanced from the more general condition of its order and class to the more particular features of its species, and finally to the singular traits of the individual. The history I have recounted thus far runs hard against this opinion. But even if we did not have the evidence of Darwin's unpublished manuscripts, notebooks, essays, and letters, there would still be the penultimate chapter of the *Origin of Species*. The most straightforward construction of that text not only confirms that he embraced the principle of recapitulation but indicates why it was essential to his theory that he do so.

122. Charles Darwin to Joseph Hooker (14 December 1859), *Life and Letters,* 2: 39.

In the text of the chapter, Darwin provides the inductive evidence for what is really a deductive conclusion from his general theory. He first observes that in respect to various species members of a given larger type, "certain organs in the individual, which when mature become widely different and serve for different purposes, are in the embryo exactly alike."[123] These points of embryonic similarity—for instance, in the wing of a bat, the paddle of a porpoise, and the hand of man—could not be due to similarity of the conditions in which they would exist as adults (as Cuvier supposed), but could most easily be explained as the heritage from a common ancestor. Indeed, the difficult classificatory problem of systematically locating barnacles (which Cuvier relegated on the basis of their adult form to the molluscan type) disappeared when their free-swimming larvae were found similar to crustaceans, their apparent ancestral forebears.

In order to account for the preservation of embryonic similarity, Darwin advanced two specific principles of adaptation, conforming to the requirements of his mechanisms of natural selection and of use and disuse. We might call them "principles of terminal addition." He specified them to be: (1) that the successive modifications by which species have attained their structure "supervened at a not very early period of life" and (2) that variations occurring in the adult recurred at a "corresponding age in the offspring."[124] These two principles, he argued, were confirmed by the practices of domestic breeders, who formed their varieties usually by selecting variations of the mature animal. The principles also received support, he maintained, from his own measurement experiments conducted on the neonates of different varieties of dogs, horses, and pigeons. These experiments showed that newborns of allied varieties were more similar to each other than to their own adult forms, and more similar than the adults to each other. If modifications supervened on the adults, embryonic forms of allied species would preserve

123. Darwin, *Origin of Species,* p. 439.
124. Darwin, *Origin of Species,* p. 444.

the similarities that these experiments revealed. More generally, however, Darwin's two mechanisms of adaptation—
natural selection, and use and disuse—required an animal to
be subjected to the pressures of life in a variegated world,
pressures which would increase in the mature organism. (Ex-
ceptions to the two principles would occur, as Darwin noted,
if the young led an independent life.)[125] For Darwin, then,
these principles of adaptation meant that generally speaking
evolutionary development proceeded by terminal additions:
new modifications would be added to the more mature animal, either through natural selection or through use and disuse, whereas "the young will remain unmodified, or be
modified in a lesser degree."[126]

125. In the first edition of the *Origin of Species,* Darwin gave examples of
young, usually the larvae of insects, that would be subject to the vagaries of
the external environment and so would acquire adaptations that would not
have characterized their adult ancestors. Through the fourth, fifth, and sixth
editions of the *Origin,* he added more examples of young animals that
"might come to pass through stages of development, perfectly distinct from
the primordial condition of their adult progenitors." See Charles Darwin,
The Origin of Species by Charles Darwin: A Variorum Text, ed. Morse
Peckham (Philadelphia: University of Pennsylvania Press, 1959), p. 700. He
nonetheless emphasized in these later editions more strongly than before
that "it is highly probable that with many animals the embryonic or larval
stages show us, more or less completely, the condition of the progenitor of
the whole group in its adult state." He then enumerated confirming examples drawn from crustacean groups and vertebrates. See ibid., pp. 702–3, and
my discussion below.

126. Darwin, *Origin of Species,* p. 447. The principle of terminal addition
for adaptive traits was already vaguely present in the Essays of 1842 (*Foundations of the Origin of Species,* pp. 42–43) and 1844 (ibid., p. 230), but was
insisted upon in the *Origin, Descent of Man* (see below), and *Variation of
Animals and Plants under Domestication.* In that latter book Darwin introduced his topic by summarizing the general features of his theory, which included: "It can be shown that modifications of structure are generally
inherited by the offspring at the same age at which each successive variation
appeared in the parents; it can further be shown that variations do not commonly supervene at a very early period of embryonic growth, and on these
two principles we can understand that most wonderful fact in the whole cir-
cuit of natural history, namely, the close similarity of the embryos within the

As a result of species descent and the operation of the two principles of adaptational inheritance, the embryo must go through the stages of its progenitors. This is a conclusion that Darwin expressly drew, even if in a somewhat tautological way: "As the embryonic state of each species and group of species partially shows us the structure of their less modified ancient progenitors, we can clearly see why ancient and extinct forms of life should resemble the embryos of their descendants,—our existing species."[127] In the text, Darwin immediately identifies this relation of the embryo to its extinct, more primitive ancestors with Agassiz's view, remarking that "Agassiz believes this to be a law of nature; but I am bound to confess that I only hope to see the law hereafter proved true."[128] Darwin realized that certain factors would inhibit the recognition of Agassiz's law: adult variations would in the long course of time be pushed back ever further into early development; as well, in certain animals variations might supervene at an early age. And of course an imperfect geological record would retard recognition of parallel development. Despite these camouflaging factors, Darwin remained sanguine that the principle would be empirically confirmed, since "this doctrine of Agassiz accords well with the theory of natural selection."[129] But even if it were not completely verified, it could, he argued, still be true.[130] His

---

same great class—for instance, those of mammals, birds, reptiles, and fish." See Charles Darwin, *The Variation of Animals and Plants under Domestication,* 2d ed. (New York: D. Appleton, 1899), 1: 12.

127. Darwin *Origin of Species,* p. 449.

128. Ibid.

129. Ibid., p. 338.

130. Despite the difficulties of securing empirical evidence, Darwin insisted that Agassiz's principle could still be true, for it was really a consequence of the theory of evolution by natural selection (ibid., p. 450): "It should also be borne in mind that the supposed law of resemblance of ancient forms of life to the embryonic stages of recent forms, may be true, but yet, owing to the geological record not extending far enough back in time, may remain for a long period or for ever, incapable of demonstration." See also text to the preceding note.

confidence that the principle was true—indeed, had to be true—even if the empirical evidence did not yet clearly demonstrate it can be measured by emendations that he introduced into the fourth (1866) and sixth (1872) editions of the *Origin.*

In the fourth edition, Darwin removed the qualification that the embryo "partially" shows the structure of the ancient progenitors; the passage became: the embryonic state "shows us more or less completely" the structure of the progenitors.[131] And by the sixth edition the passage made clear that the embryo traversed the "adult" stages of the progenitors: "As the embryo often shows us more or less plainly the structure of the less modified and ancient progenitor of the group, we can see why ancient and extinct forms so often resemble in their adult state the embryos of existing species of the same class."[132] In the sixth edition Darwin also expressed more firmly his confidence in Agassiz's law: "Agassiz believes this to be a universal law of nature; and we may hope hereafter to see the law proved true."[133]

Darwin preserved his hope that empirical evidence would clearly demonstrate Agassiz's law, since the law itself was a deductive conclusion from the general theory of evolution by natural selection. Evidence might be accumulated from the source that Agassiz originally mined, namely, the paleontological record. But here Darwin ran into large difficulties. Geological deposits should record, if Darwin's theory was correct, a gradual transition of forms into ever more progressive types. But fossil deposits show large gaps between forms, with only some vague indications of advancing development. Geological remains actually accorded more with Agassiz's notion of abrupt catastrophes and creative repopulations through the vast eras of time. In chapters 9 and 10 of the *Origin,* Darwin expended his ingenuity in explain-

131. Darwin, *Origin of Species: Variorum Text,* p. 704.
132. Ibid.
133. Ibid.

ing why the geological record was imperfect and indeed had
to be. We should expect an incomplete record because, as
Darwin enumerated the causes: few areas of the earth's sur-
face have been systematically explored; soft-bodied organisms would not fossilize; bones and shells would best be
preserved only where sediment accumulated over them, but
there would be few of these places; successive geological formations, apparently temporally contiguous, were actually
separated by enormous periods of time; erosion would destroy many deposits; uplift and subsidence of land would dis-
rupt patterns of deposition; and animals migrating into new
territories would have no predecessors already entombed in
the rocks. And if the theory of evolution was correct, we
should expect transitional forms to be preserved in strata far
below their divergent lineages; and in speciating groups,
competitive exclusion would greatly reduce the sheer number of the parent types.[134] For all of these reasons the book of
life must be as the rat-chewed tomes of an abandoned monas-
tery library. "Of this volume," Darwin reflected, "only here
and there a short chapter has been preserved; and of each
page, only here and there a few lines."[135]

Darwin had explained why paleontology did not yield up
the story of life's gradual evolution. But his account was really
too good: for the reasons he enumerated, geology alone
could never supply convincing evidence for this theory. Yet,
Darwin suggested, there was way of filling in the missing
pages of life's progressive history, and thus the geological evidence could be made legible. The fragments of the story
might be connected through the picture left by the embryo as
it passed through the various stages of its developmental advance. At the conclusion of his two chapters on the imperfec-
tions of the geological record, Darwin linked the scattered
evidence of progress found by paleontologists, like Agassiz,
with his theory of embryological recapitulation. Recapitula-

134. Darwin, *Origin or Species,* chaps. 9 and 10.
135. Ibid., pp. 310–11.

tion would show that the intimations of gradual, progressive evolution left in the rocks were no lusus naturae. Indeed, according to Darwin, recapitulation would make comprehensible how natural selection could be the agent of progressive evolution, since it accounted for both the embryo's development and that gradual improvement hinted at in the tattered book of life.

> The inhabitants of each successive period in the world's history have beaten their predecessors in the race for life, and are, in so far, higher in the scale of nature; and this may account for that vague yet ill defined sentiment, felt by many palaeontologists, that organisation on the whole has progressed. If it should hereafter be proved that ancient animals resemble to a certain extent the embryos of more recent animals of the same class, the fact will be intelligible.[136]

The principle of recapitulation thus might help recover those missing pages of the geological record and so produce strong empirical confirmation of the truth of evolution by natural selection. During the period leading up to the composition of the *Descent of Man,* Darwin looked to recapitulation precisely as providing the empirical support he sought.

The Role of Recapitulation in the Descent of Man

Several events occurred during the decade of the 1860s to bolster Darwin's faith in recapitulation theory. First, important authorities came to endorse the position he had set out in chapter 13 of the *Origin of Species.* In 1864 Darwin received a little book, *Für Darwin* (1863), from Fritz Müller (1821–97), a South American naturalist with whom he then initiated a lifelong correspondence. He was so impressed with the book, especially with the chapters on classification and embryology, that he arranged to have it translated.[137] It ap-

136. Ibid., p. 345.
137. Charles Darwin to Fritz Müller (10 August 1865), *Life and Letters of Charles Darwin,* 2: 222.

peared as *Facts and Arguments for Darwin* in 1869. In the book Müller fostered and developed Darwin's ideas about recapitulation. Not surprisingly, perhaps, it was Ernst Haeckel who first recommend Müller's work to Darwin;[138] and Haeckel looked to both Darwin and Müller as the modern originators of the principle he further developed as the "biogenetic law." Certainly Haeckel could open the sixth edition of the *Origin of Species* to find the discoverers of modern recapitulation in full embrace. For there Darwin argued, now following Müller, that since the larvae of quite diverse species of crustaceans displayed the same characteristic nauplius form, "it is probable that an independent adult animal, resembling the nauplius, formerly existed at a remote period, and has subsequently produced, through long-continued modification along several divergent lines of descent, the several above-named great Crustacean groups."[139] And Darwin quite immediately generalized such evidence as that provided by crustacean larvae: "So again it is probable, from what we know of the embryos of mammals, birds, fishes, and reptiles, that these animals are the modified descendants of some ancient progenitor, which was furnished in its adult state with brachiae, a swim-bladder, four fin-like limbs, and a long tail, all fitted for an aquatic life."[140] Müller and Haeckel thus confirmed the principle of recapitulation for Darwin and stimulated him to give it a more unqualified expression in subsequent editions of the *Origin*. Yet despite the support he had received from his German colleagues, he still ardently desired to win Huxley over.

Through the early 1860s, Huxley began to soften. Darwin initially was quite apprehensive about what his friend would say of his embryological ideas in the *Origin*. Writing Joseph Hooker in December 1859, he added the postscript:

> I shall be very curious to hear what you think of my discussion on Classification in Chap. XIII.; I believe

138. Ernst Haeckel to Charles Darwin (26 October 1864), DAR 166.1.
139. Darwin, *Origin of Species: Variorum Text,* p. 702.
140. Ibid.

> Huxley demurs to the whole, and says he has nailed
> his colours to the mast, and I would sooner die than
> give up; so that we are in as fine a frame of mind to
> discuss the point as any two religionists. Embryology
> is my pet bit in my book, and, confound my friends,
> not one has noticed this to me.[141]

Two months later, on 10 February 1860, Darwin's anxiety was compounded as a result of Huxley's lecture on evolutionary theory at the Royal Institution. Darwin was disappointed because his "general agent"[142] hesitated about whether natural selection could produce sterility between species, and especially because, as he wrote to Hooker, "Huxley has never alluded to my explanation of classification, morphology, embryology, etc."[143] Darwin thought Huxley was quite dissatisfied with these parts of his theory. Private conversation, however, apparently persuaded him otherwise. He continued in his letter: "But to my joy I find it is not so, and that he agrees with my manner of looking at the subject."[144]

After a fashion Huxley did come around. In his little book of 1863, *Man's Place in Nature,* he presented the morphological evidence for Darwin's theory, including the embryological. Using illustrations of embryos of a dog and of a human being to demonstrate structural similarity, Huxley concluded with a remark covering a neutral ground between von Baer and Darwin: "Without question, the mode of origin and the early stages of the development of man are identical with those of the animals immediately below him."[145] Since Huxley did not mention the crucial part of the von Baerean

141. Charles Darwin to Joseph Hooker (14 December 1859), *Life and Letters of Charles Darwin,* 2: 39.

142. It was at this time that Darwin started referring to Huxley as his "general agent." See Francis Darwin's note in *Life and Letters of Charles Darwin,* 2: 46.

143. Charles Darwin to Joseph Hooker (14 March 1860), *More Letters of Charles Darwin,* 2: 138.

144. Ibid.

145. Thomas Henry Huxley, *Evidence as to Man's Place in Nature* (London: Williams and Norgate, 1863; photoreproduced by University of Michigan Press, 1959), p. 81.

principle—namely, that embryological development merely moved from the abstract, universal characteristics of a type to its particular embodiment—he did not rule out Darwin's interpretation; but since he failed to suggest that the Urtype had been embodied in a real animal, he did not exactly endorse Darwin's interpretation either, nor would he for some long time.[146] It was safe ground for a general agent with a scientific conscience.

Qui tacet consentire, as the law argues. Darwin took Huxley's silence in the 1860s as acquiescence. In the *Descent of Man* (1871), he also employed woodcuts of comparable embryonic states in a dog and human being (the originals from German sources were recommended, he noted, by Huxley and Haeckel) to illustrate similarity of development (see fig. 30), citing as support the passage from Huxley that I have just quoted. But then he went further than the doubtful Huxley of 1863 would have liked. After mentioning agreement by authorities, principally Huxley, on parallel development of embryonic states, he cautiously slipped in: "It may, however, be added that the human embryo likewise resembles in various points of structure certain low forms when adult."[147] He then quickly illustrated the recapitulation thesis with examples of organs emerging in the human embryo that

146. Through the 1860s, Huxley seems to have appreciated the potency of Darwin's arguments for recapitulation, yet his commitment to von Baer's embryology restrained a public reversal of his earlier-stated opinions. Even his long review of Haeckel's *Naturliche Schöpfungsgeschichte* at decade's end remained silent on the issue: indeed, nowhere in the review did he even mention recapitulation, the chief principle of Haeckel's evolutionary theory. Only in 1878 did Huxley admit that, properly interpreted, Meckel's views on recapitulation did conform to acceptable evolutionary doctrine. See Thomas Huxley, "The Natural History of Creation," *Academy* 1 (1869): 13–14, 40–43; and "Evolution," *Encyclopaedia Britannica,* 9th ed. (1878), 8: 744–51. In his encyclopedia article Huxley conceded: "the reptile embryo, at one stage of its development, is an organism which, if it had an independent existence, must be classified among fishes, and all the organs of the reptile pass, in the course of their development, through conditions which are closely analogous to those which are permanent in some fishes" (ibid., p. 750).

147. Charles Darwin, *The Descent of Man and Selection in Relation to Sex* (London: Murray, 1871), 1: 16.

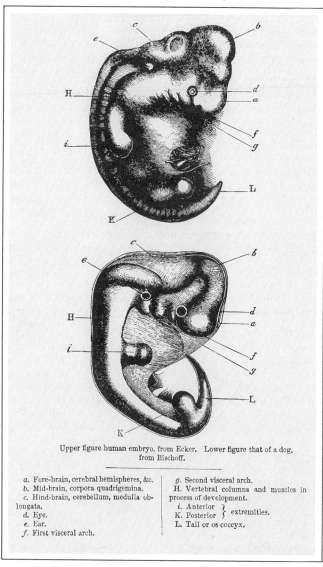

Upper figure human embryo, from Ecker. Lower figure that of a dog, from Bischoff.

<table>
<tr><td>a. Fore-brain, cerebral hemispheres, &c.
b. Mid-brain, corpora quadrigemina.
c. Hind-brain, cerebellum, medulla ob-
longata.
d. Eye.
e. Ear.
f. First visceral arch.</td><td>g. Second visceral arch.
H. Vertebral columns and muscles in
process of development.
i. Anterior }
K. Posterior } extremities.
L. Tail or os coccyx.</td></tr>
</table>

FIGURE 30. Illustration of homologies of human and dog embryos, from Charles Darwin, *Descent of Man* (1871).

had their counterparts in the ascending phylogenetic sequence of lower adult organisms—so, for instance, he cited Theodor Bischoff's observation that the brain of the human fetus in the seventh month reaches "about the same stage of development as in a baboon when adult."[148] Darwin justified his evolutionary recapitulationism as he had in the *Origin,* namely, by his principle of terminal adaptational addition, that is, "the principle of variations supervening at a rather late embryonic period."[149]

The real payoff for the principle of embryological recapitulation came, however, later in the *Descent.* Though Huxley admitted embryological similarity (suggesting a real relation of descent) within particular archetypes—the vertebrata, for instance—he was mute about embryological relations that might indicate transitions across archetypes. While still in ardent thrall to von Baer in the early part of the 1860s, Huxley could not admit a gradual transition from one type to another (hence his early notions about evolutionary saltations).[150] But gradual, progressive transitions formed the heart of Darwin's descent theory. Darwin received from an unexpected source just the kind of embryological evidence he needed to further solidify a general use of the principle of recapitulation and to reach one of those Archimedean factual points necessary to raise the theory of gradual, progressive descent to greater probability. The Russian paleontologist Vladimir Kovalevsky wrote in March 1867 to ask permission to translate the *Variation of Plants and Animals under Domestication;* serendipitously he sent along some articles by his brother Alexander that Darwin found most interesting.[151] In those articles and in subsequent correspon-

---

148. Ibid., p. 16.

149. Ibid., p. 32.

150. Lyons argues that Huxley's saltationism derived from his earlier commitment to von Baer and the theory of distinct Cuvierian types. See her "The Evolution of Thomas Henry Huxley's Evolutionary Views."

151. Vladimir Kovalevsky to Charles Darwin (15 March 1867), DAR 169. See especially A. Kovalevsky, "Entwickelungsgeschichte der einfachen Ascidien,"

dence,[152] Kovalevsky attempted to demonstrate that larval ascidians, classified as invertebrates, were morphologically similar to amphioxus, a primitive vertebrate. From the evidence provided by Kovalesky of the several traits of similar structure—especially the axial cylinder in the tail of the larval ascidian and the chorda dorsalis of amphioxus[153]—Darwin drew a singularly important conclusion for his theory:

> We should thus be justified in believing that at an extremely remote period a group of animals existed, resembling in many respects the larvae of our present Ascidians, which diverged into two great branches—the one retrograding in development and producing the present class of Ascidians, the other rising to the crown and summit of the animal kingdom by giving birth to the Vertebrata.[154]

Wielding the recapitulation principle, Darwin thus suggested that the invertebrate and vertebrate types had been evolutionarily united through the ascidians. As this passage also indicates, he believed that on the basis of embryological evolution he could reconstruct phylogenetic history. (And actually he employed this same methodological use of the principle of recapitulation even earlier, when he relocated barnacles in the class of crustaceans because of the morphological similarity between their free-swimming larvae and adult crustaceans.) When Ernst Haeckel similarly used the biogenetic law to reconstruct phylogenies, he was certainly working within the *echt* Darwinian tradition.

Mémoires de l'Académie impériale des sciences de St.-Pétersbourg, 7th ser., vol. 10, no. 15 (1866). Because his findings were challenged, Kovalesky followed up his original study with another article, which he sent to Darwin just before the publication of the *Descent.* See A. Kovalevsky, "Weitere Studien über die Entwicklung der einfachen Ascidien," *Archiv für Mikroskopische Anatomie* 7 (1871): 101–30.

152. Alexander Kovalevsky to Charles Darwin (25 September 1870), DAR 169.

153. A. Kovalesky, "Entwickelungsgeschichte der einfachen Ascidien," p. 13.

154. Darwin, *Descent of Man,* 1: 206.

The Logic of Darwin's Theory of Evolution

In the *Origin of Species* and the *Descent of Man* Darwin gathered the empirical evidence that supported his theory of embryological recapitulation and therewith unequivocally expressed his belief in its truth. But it was not simply a few facts that convinced him. As I have tried to show by the slow delivery of this history, the principle of embryological recapitulation was woven into the original fabric of Darwin's conception of evolution. By 1859 the principle had become a deductive consequence. It is important to understand this logical relationship in order to evaluate the claims of some historians who believe they can distinguish a true Darwinian theory of evolution from miscreant theories having embryological development at their core. The necessity, for Darwin, of the principle of recapitulation can be understood, I believe, when we directly consider what "descent with modification" meant for him and how it solved for him the great questions concerning "the unity of type."

Darwin believed that "the grand fact in natural history," as he called it, "of the subordination of group under group," or "the unity of type," had been explained by his theory of descent.[155] Subordination of orders to a class, of families to an order, of genera to a family, and finally of species to a genus would become comprehensible if we supposed that the homologous resemblances that snaked through the classificational tree were genealogical. Thus the similarity of species within a genus could be naturally explained if in the long past the genus itself had been a species that produced varieties, which then gradually evolved into the species presently grouped under that genus. Now Darwin did come, as Gould and Ospovat believe, to accept von Baer's law that the embryo develops from a more generalized condition of its type to the more particularized features of its species. Owen, Barry, and Carpenter beat that idea into his early notion of how, in the words of his first Transmutation Notebook, "genera would be formed—bearing relation to ancient types." The pith of his

155. Darwin, *Origin of Species,* pp. 413, 434–35.

theory, then, was simply that the generalized set of characters that determined a type—whether the type constituted the genus, family, order, or class—was originally embodied in a population of real creatures which therefore would share common, that is, archetypal, features with their descendants. As Darwin incisively sketched this position in the *Origin:* "If we suppose that the ancient progenitor, the archetype as it may be called, of all mammals, had its limbs constructed on the existing general pattern, for whatever purpose they served, we can at once perceive the plain signification of the homologous construction of the limbs throughout the whole class."[156] The archetype is the ancestor. But if you add to von Baer's law Darwin's theory of adaptation (in which new traits are joined to the terminal, that is, more adult phases of development) and situate both in the framework of descent (in which the archetype is the ancestor), then it follows that ideally *the embryo must recapitulate* the ancient forms of its adult progenitors. Looking now to the future, as the embryo serves as the progressively developing platform for additional modifications, the phylogenetic tree will branch with new species growth. Thus embryological recapitulation becomes an exact and necessary deduction from Darwin's theory and von Baer's law. So contrary to the received historical opinion, Darwin could not, logically could not, accept both von Baer's law and his own conceptions of descent and adaptation but reject recapitulation. And Darwin understood this perfectly.

156. Ibid., p. 435.

6

THE MEANING OF
EVOLUTION AND THE
IDEOLOGICAL USES
OF HISTORY

The term "evolution" is pregnant with its history. The word had been appropriated in the seventeenth century to refer to Swammerdam's theory of the preformation of the adult within the embryo. Opposition grew to evolutionary preformationism, particularly in the versions of the leading proponents, Haller and Bonnet. Wolff's careful microscopical studies and rich arguments, especially for the role of independent formative forces, brought epigenetical theories of development to prominence. Under this intellectual pressure, the concept of the unrolling or evolution of the embryo bent toward theories of gradual development. By the later part of the eighteenth century, Autenrieth was using *evolutio* to describe the growth of the embryo, not sparked immediately into its adult form, but gradually changing through the permanent forms of lower species until the adult form was achieved. The thesis that the embryo recapitulated the morphological types of those species below it became a directive idea for a number of Germany's and France's leading biological researchers—Oken, Treviranus, Tiedemann, Meckel, and Serres. And most of these, except for Oken and Serres, were led through the notion of embryological evolution to that of species alteration: they came to believe that the embryo traversed the forms that species went through in their gradual evolutionary transformation.

Von Baer condemned both of these theories of "evolution," that of the embryo and that of the species, and did so under the common name. Of course there was nothing canonical in the term itself. Von Baer and other Germans would alternatively refer to the *Entwickelung* of embryo and species, just as English speakers would as often use "development," or sometimes "transformation" and "transmutation." Indeed, Darwin himself used a form of the word "evolution" only once in the *Origin of Species;*[1] though with the currency supplied by Spencer, he later employed it frequently to refer to his theory. The more important linguistic phenomenon to recognize, though, is that these several terms were used for both embryological and species change. Thus, while Lyell, Grant, Green, Owen, Chambers, and Carpenter spoke of "evolution," as well as of "development" and "transformation," what emerges as significant for an index of underlying concepts is that they used one term to describe the two processes. For von Baer, writing in the 1820s, that usage was conceptually appropriate; for he understood embryological evolution and species evolution to be two expressions of the same, underlying idea of morphological transition. In the next decade, Owen, under the rubric of evolution, also conceived both to be really the same supposed process; and like von Baer, he attacked both for the same reasons. Darwin, I believe, also regarded them as virtually the same process, but defended them and explained them through the same mechanisms of heritable modification—that is, through natural selection, and use and disuse. Because of the fundamental identity of these two modes of development, Darwin could

1. Charles Darwin, *On the Origin of Species* (London: Murray, 1959), p. 490. Darwin honed this lyrical passage through the Essays of 1842 and 1844. It draws the *Origin* to a close: "Thus, from the war of nature, from famine and death, the most exalted object which we are capable of conceiving, namely, the production of the higher animals, directly follows. There is grandeur in this view of life, with its several powers, having been originally breathed into a few forms or into one; and that, whilst this planet has gone cycling on according to the fixed law of gravity, from so simple a beginning endless forms most beautiful and most wonderful have been, and are being, evolved."

employ embryological evolution as a model for species evolution. And indeed, for Darwin embryological evolution became part of the casual matrix that produced species evolution, or so has been a thesis of this book.

Most recent historians of science, though, think otherwise. They maintain that Darwin had the better sense to reject recapitulation theory and instead adopt von Baer's "law" that the embryonic form simply changed from a more homogeneous and universal condition of the type to a more heterogeneous and special condition of the species and individual. This means, according to the common interpretation, that Darwinian evolutionary theory must be distinguished from those transformational conceptions that did employ an embryological model. Yet Darwin's notebooks, essays, unpublished manuscripts, and letters provide a very different historical conclusion. They place his theory, for all its distinctive features and demonstrable general validity, in the common context—or at least so I have argued. But even if we put aside that accumulated evidence, there remains the plain meaning of the penultimate chapter of the *Origin of Species*. At first glance it is extremely difficult to understand how the received historical interpretation could yet remain unabashed in view of the naked meaning of such text. The historian who provides the most instruction on this puzzle is Gould, who in his *Ontogeny and Phylogeny* drapes the relevant passages from the *Origin* with diverting finery.

At the beginning of his pages on Darwin's embryology, Gould posts the claim which guides much of the historical analysis in the rest of the book: "Darwin had accepted," he contends, "the observation of von Baer—a flat denial of recapitulation."[2] What crucially separates Darwin from the recapitulationists, he believes, is that Darwin denied any "repetition of *adult* stages in ontogeny."[3] Gould, with charac-

2. Stephen Jay Gould, *Ontogeny and Phylogeny* (Cambridge, Mass.: Harvard University Press, 1977), p. 70.

3. Ibid., p. 70, note.

teristic insight, appreciates how Darwin transformed von Baer's law so that the generalized state of the embryo was to be understood as the "bauplan . . . of a common ancestor."[4] This indeed meant, as Gould quotes Darwin, that "community in embryonic structure reveals community of descent."[5] But now comes, in his reconstruction, the apparently straightforward passage in the first edition of the *Origin* that in the previous chapter I have traced through its several emendations in subsequent editions.[6] After quoting that passage and a comparable one from the 1844 Essay, Gould asks: "But is not Darwin perilously close to recapitulation at this point?" Well close, perhaps, but he thinks Darwin did not capitulate. Gould has two general grounds for distinguishing Darwinian evolutionary embryology from that of Fritz Müller and Ernst Haeckel. The first rationale he offers is that "the two views imply radically different concepts of variation, heredity, and adaptation." The second is secured by a subtle reading of Darwin:

> Darwin saw that ancestral groups in an established community of descent would differ least in their adult form from the embryonic state common to all members of the community. The gill slits of the human fetus represent no ancestral adult fish: we see no repetition of adult stages, no recapitulation. Yet adult fish, as primitive ancestors, have departed least from this embryological condition of all vertebrates.[7]

Let us first consider this reading. Darwin certainly held that the ancient adult progenitor of man was morphologically closer to the vertebrate archetype and also to its own embryo. But as I have tried to indicate in the preceding chapter, for

4. Ibid., p. 72.

5. Darwin, *Origin of Species,* p. 449. Gould (*Ontogeny and Phylogeny,* p. 72) believes this "must rank as Darwin's primary statement of the relationship between embryology and evolution."

6. See text to n. 127 of chap. 5.

7. Gould, *Ontogeny and Phylogeny,* p. 72.

Darwin this *did* mean that the human embryo passed through the adult stage of its ancestors—not only was this for Darwin a logical consequence of the situation (that is, if the adult ancestor and its embryo did not morphologically differ, they must have represented the same form) but Darwin actually said it was the "adult" stage in later editions of the *Origin* and in the *Descent.* But there is a deeper reason, of which Darwin was fully cognizant, for the requirement that the embryo pass through the adult stages of its predecessors. Gill slits, heterocercal tail, and any other traits possessed by ancient progenitors could not have been the original endowments of animals "that have departed least from this embryological condition of all the vertebrates." Darwin was not a creationist. These also had to be adaptations acquired in the course of evolution. And according to Darwin's constantly reiterated principle that such modifications would generally be acquired by more mature or adult organisms, these traits and all others must have supervened upon the adults of a yet-earlier class of progenitors. For Darwin it was adults all the way down— when embryos manifested traits acquired by progenitors, it had to be the adult traits they manifested. Embryos thus had to pass through the *adult* stages of their ancestors.

Darwin's theories of adaptation and heredity, by which common traits would come to invest both ancestor and descendent embryo, did not differ significantly from those of Fritz Müller and Ernst Haeckel, both of whom looked to Darwin as the originator of these ideas. In the embryological sections of the Essays and *Origin,* in the *Variation of Animals and Plants,* and in the *Descent of Man,* as well as in Darwin's scattered notes, he urged the principle of what I have called, adopting Gould's language, "terminal adaptational additions"—the hypothesis that the modifications by which species have attained their structures have generally "supervened at a not very early period of life."[8] This is precisely the

---

8. Darwin, *Origin of Species,* p. 444. See my discussion in the last three sections of chap. 5.

principle that Gould attacks when Haeckel used it for the
same theoretical purpose.[9] Gould thinks that the principle is
bound to the Lamarckian idea of the inheritance of acquired
characteristics, which Haeckel certainly endorsed. This connection, then, condemns the principle of terminal addition
and ultimately recapitulation theory in neo-Darwinian eyes.
But of course Darwin was not a neo-Darwinian. He too defended "Lamarckian inheritance." Indeed, in the only place in
the embryological section of the *Origin* in which he expressly talks about the mechanism of heritable acquisition, it
is "use and disuse" to which he appeals[10]—though natural selection could just as well provide terminal additions, and in
other of his writings he maintains that it does. As to the consequence of terminal additions, Darwin drew the direct
conclusion—it led, as he expressly declared in the first edition of the *Origin,* to embryonic recapitulation of adult forms:

> the adult differs from its embryo, owing to variations
> supervening at a not early age, and being inherited at
> a corresponding age. This process, whilst it leaves
> the embryo almost unaltered, continually adds, in
> the course of successive generations, more and
> more difference to the adult. Thus the embryo
> comes to be left as a sort of picture, preserved by na-
> ture, of the ancient and less modified condition of
> each animal.[11]

Like Haeckel, Darwin clearly viewed the development of the
embryo as comparable to flipping through a series of da-

9. Gould, *Ontogeny and Phylogeny,* p. 81.

10. Darwin, *Origin of Species,* p. 447: "Whatever influence long-continued
exercise or use on the one hand, and disuse on the other, may have in modi-
fying an organ, such influence will mainly affect the mature animal, which
has come to its full powers of activity and has to gain its own living; and the
effects thus produced will be inherited at a corresponding mature age.
Whereas the young will remain unmodified, or be modified in a lesser degree by the effects of use and disuse."

11. Ibid., p. 338.

FIGURE 31. Ernst Haeckel (seated), 1834–1919, who gave currency to the biogenetic law that ontogeny recapitulated phylogeny; photograph taken in 1866.

guerreotypes, which would produce a dynamic picture of the phylogenetic history of the species.

To reconstruct Darwin's theory, then, as denying recapitulation must strain against the weight of plain assertions, easily interpreted, and logical requirements, easily understood. What, then, explains the efforts of Gould, Mayr, and Bowler (just to mention the most prominent and influential historians on this question)? I think it can only be ideology.

The past has always served the ideological interests of political and social thinkers. It confers on their ideas respect, authority, and the wisdom of age. Historians and scientists as well have marshaled the forces of the past for these same purposes. When answering in this way, though, I do not mean to suggest that a vigilant historian or scientist could avoid refracting his or her subject through a dark glass of obscure and virtually unconscious assumptions. I am sure I am not able to do so. My own narrative undoubtedly depends on some considerations that bend this history in ways that would appear distorting to others. But we can all be helped by friends and critical readers, so let me suggest a social and scientific reconstruction of this historiography. But first a few words about the use of the term "ideology."

To call a set of ideas "ideological" and the one holding them an "ideologue" would have been a neutral assessment at the end of the eighteenth century. Just after the French Revolution, Destutt de Tracy proclaimed a science of ideas that would have as its special concern those conditions that vanished into the very fabric of the mind. He wished to uncover their source and test their validity against the firm base of sensation. Marx emphasized one aspect of Destutt's doctrine, namely, the discovery of unfounded conceptions. He used "ideology" to refer to the class of ideas that conflicted with science, particularly the scientific analysis of history—ideas that were irrational, scientifically groundless, and usually determined by the special interests of a group. Ideology thus evolved from a science of all ideas to become a category describing ideas that have no scientific foundation. The safe use of the term "ideology," however, supposes one can distinguish an authentic science or history from those that are epistemologically tainted. But the efforts of the first half of this century to ground science and history on solid fact and true theory have, like chimney sweeps, come to dust. These failures of foundationalism have clouded any other distinctions that might yet even tentatively separate good history and science from ideologically perfused enterprises. Robert Young thus feels no compunctions in calling science in general and Darwinian theory in particular "ideological."[12] And so the social constructionists have taken heart.

I think, however, we can still use the term "ideological" without suggesting that all science and history are contaminated, that they are no more epistemologically secure than football (i.e., different game, same stakes). When I call the current historiography concerning the issues treated in this book "ideological," I mean that such historical work meets four criteria, which I wish to suggest for a reasonable deployment of the term.

A historical representation will be ideological (to greater or lesser degree), then, if the following conditions obtain: first, the historical account employs an interpretative framework or set of assumptions that are covert and neither justified nor argued for in the account; second, the framework or assumptions express the shared values and position of a particular community rather than the idiosyncratic view of the historian; third, the main function of the framework or assumptions is to justify the shared values and position rather than to realize the principal value of recovering the past; and finally, the historian's interpretations and arguments serve chiefly to justify the framework and thus the values. The last two conditions, of course, are crucial. The critic making the charge of ideological taint will maintain that the history has been bent to serve the purpose of indirectly arguing for a

---

12. Robert Young, "Darwinism is Social," *The Darwinian Heritage,* ed. David Kohn (Princeton: Princeton University Press, 1985), pp. 609–38.

position and set of values that are independent of the effort to represent the past *wie es eigentlich gewesen*. Though the Rankean expression embodies an unattainable goal, it must yet remain the shibboleth of the tribe of historians. That a given historical treatment fails to realize the Rankean intention must itself be a historical judgment. Such a judgment, really an indictment of the account under review, must be based both on a more accurate and richer historical construction of the subject (in this case, of Darwin's positions on recapitulation and progress) and on a warranted argument for the presence in the work of a framework and values other than the Rankean.[13] Now what might be the underlying set of considerations that have shaped the historical analyses of Gould, Mayr, Bowler, and those of like twentieth-century mind?

Gould and Mayr have a scientific interest in reading Darwin as they have. He is the patron saint of evolutionary biology— and for very good reason. To have his blessing on scientific positions one wishes to maintain in the late twentieth century can only advance their cause. Both of these historian-scientists regard freely flowing variational possibilities as the juice of evolution; and suspect constraints (like recapitulation) that act to inhibit the flow can, they believe, only produce stagnation. But more fundamentally they reject utterly any notions of guidance in evolution by teleological factors, more than a whiff of which the history of recapitulation exudes. Recapitulation theory has always been joined to ideals

13. Leopold von Ranke (1795–1886) had an ultimate goal in securing the past "as it really was." He wished to trace out the divine design in the affairs of human beings. He intended his interpretative framework, as contrasted with that of Hegel, to be used in the service of capturing the past in its authentic particularity. For only by the conduct of a "scientific history" could he realize his ultimate aim of justifying God's ways to men. His success, like that of any historian, must be judged, as I have suggested, by a careful reinvestigation of the historical events recounted and by a close analysis of the recounting of those events. For a masterful discussion of Ranke and his context, see Georg Iggers, *The German Conception of History*, 2d ed. (Middletown, Conn.: Wesleyan University Press, 1983).

of progress; and for Gould and Mayr progressivist evolutionary processes can only be the result of fixed goals to be achieved—teleology in another guise. All of these unhappy scientific changelings could be more easily buried if Darwin himself were to chant the obsequies.

The ideology that influences the interpretation of the past need not itself have unsavory qualities. The contemporary scientific assumptions that Gould and Mayr read back into Darwin's thought certainly have proved fruitful in the development of vigorous neo-Darwinian theory. The ethical and social components of that ideology are not merely acceptable but quite admirable. They do, however, distort the history. Gould finds in Haeckelian evolutionary thought, with its obvious progressivism, political Prussianism, and racism, a harbinger of Nazi horrors.[14] It would therefore be simply impossible to allow Haeckel's chief biological principle, the biogenetic law that ontogeny recapitulates phylogeny, to have been also endorsed by Darwin. Social Darwinism, with all its real and apparent excesses, must have sources other than its name suggests. Bowler shares both this humane perspective and the need to certify an authentic Darwinism. Like Gould, he also finds in Haeckel's and Spencer's evolutionism the scientific font spewing "callous indifference to the fate of those individuals or races that could not keep up with the march of progress."[15] The mistake made by cultural historians, he believes, is that they have not recognized two distinct lines of nineteenth-century evolutionary thought: Darwin's theories, which were nondevelopmental, nonrecapitulational, and

14. Gould, *Ontogeny and Phylogeny,* pp. 77–78: "His [Haeckel's] evolutionary racism; his call to the German people for racial purity and unflinching devotion to a 'just' state; his belief that harsh, inexorable laws of evolution ruled human civilization and nature alike, conferring upon favored races the right to dominate others; the irrational mysticism that had always stood in strange communion with his brave words about objective science—all contributed to the rise of Nazism."

15. Peter Bowler, *The Non-Darwinian Revolution: Reinterpreting a Historical Myth* (Baltimore: Johns Hopkins University Press, 1988), p. 198.

Figure 32. Charles Darwin, portrait done in 1884.

nonprogressivist, and the theories of Haeckel and others which had the opposite features—and for Bowler it is "recapitulation theory [that] illustrates the non-Darwinian character of Haeckel's evolutionism."[16] Because cultural historians have not distinguished these two strands of biological thought, "there is a continuous process linking nineteenth-century 'Darwinism' to modern ideas of genetical determinism and sociobiology."[17] Bowler's history thus aims to cut away the deadly tangle of pseudo-Darwinian ideas from real Darwinian theory. The attempt to do this while staring directly into the face of the late twentieth century can, however, only produce dull stone. The knotty skein of Medusa's locks must be slowly and carefully coaxed to unfold, as the tedious argument of this essay may have demonstrated.

At the end of the twentieth century, over one hundred and fifty years after Darwin first formulated his theories of evolution, his conceptions have become, to paraphrase Auden's lines about Freud, a whole climate of opinion. The debates today range over the tempo and mode of evolution: whether evolution is gradual and constant or saltational and occasional; whether it is driven to great contingent branching by proliferating variability or is more narrowly confined by reduction in variability; whether transformation is directed by natural selection alone or also by the structural constraints of the organism; whether selection focuses exclusively on minimal hereditary units or also on systems of greater comprehension—from individuals through kin groups to populations and species. All of these different positions can easily be traced back through the nineteenth century, with Darwinians then disputing among themselves much as now. But neo-Darwinians seem to have reached general agreement that three older proposals should be dismissed: that species evolution should be modeled on individual evolution, that embryogenesis recapitulates phylogenesis, and that

16. Ibid., p. 84.
17. Ibid., p. 199.

evolution is progressive. It is thus surprising to discover that these ideas nonetheless served in the *Bauplan* of Darwin's thought. Darwin was indeed the architect of the theory that has been reconstructed as neo-Darwinism. But the architect was our ancestor, who dwelt happily enough in the nine-teenth century.

BIBLIOGRAPHY

Manuscript Collection: Darwin Archive, Cambridge University Library.

Agassiz, Louis. *Histoire naturelle des poissons d'eau douce.* 3 vols. Neuchâtel: Petitpierre, 1842.

———. *Lake Superior: Its Physical Character, Vegetation, and Animals.* Boston: Gould, Kendall and Lincoln, 1850.

———. "A Period in the History of our Planet." *Edinburgh New Philosophical Journal* 35 (1843): 1–27.

Aristotle. *The Complete Works of Aristotle: The Revised Oxford translation.* Edited by Jonathan Barnes. 2 vols. Princeton: Princeton University Press, 1984.

Appel, Toby. *The Cuvier-Geoffroy Debate.* Oxford: Oxford University Press, 1987.

Autenrieth, Johann Heinrich. *Supplementa ad historiam embryonis humani.* Tübingen: Heerbrandt, 1797.

Bacon, Francis. *Sylva Sylvarum: Or a Naturell Historie in Ten Centuries.* 3d ed. London: William Lee, 1631.

Baer, Karl Ernst von. *Autobiography of Dr. Karl Ernst von Baer.* Edited by Jane Oppenheimer. Translated by H. Schneider. Canton, Mass.: Science History Publications, 1986.

———. *Entwickelungsgeschichte der Thiere: Beobachtung und Reflexion.* Königsberg: Bornträger, 1828.

———. "Fragments relating to Philosophical Zoology: Selections from the Works of K. E. von Baer." Translated by Thomas Henry Huxley. In *Scientific Memoirs, Selected from the Transactions of Foreign Academies of Science, and from Foreign Journals: Natural History,* edited by Arthur Henfrey and Thomas Henry Huxley. London: Taylor and Francis, 1853.

Barry, Martin. "Further Observations on the Unity of Structure in the Animal Kingdom." *Edinburgh New Philosophical Journal* 22 (1836–37): 345–64.

———. "On the Unity of Structure in the Animal Kingdom." *Edinburgh New Philosophical Journal* 22 (1836–37): 116–41.

Blumenbach, Johann Friedrich. *Handbuch der Naturgeschichte.* 12th ed. Göttingen: Dieterich'schen Buchhandlung, 1830.

Boerhaave, Hermann. *Praelectiones academicae.* 6 vols. Edited with notes by Albertus Haller. Göttingen: Vandenhoeck, 1744–48.

Bonnet, Charles. *Considerations sur les corps organisés.* 2 vols. Amsterdam: Marc-Michel Rey, 1762.

———. *La Palingénésie philosophique, ou Idées sur l'état passé et sur l'état futur des êtres vivans.* 2 vols. Geneva: Philibert et Chiroi, 1769.

Bowler, Peter. "The Changing Meaning of 'Evolution.'" *Journal of the History of Ideas* 36 (1975): 95–114.

———. *The Non-Darwinian Revolution: Reinterpreting a Historical Myth.* Baltimore: Johns Hopkins University Press, 1988.

Buffon, Georges Louis Leclerc, Comte de. *Oeuvres complètes de Buffon.* Edited by Pierre Flourens. 12 vols. Paris: Garnier, 1852–55.

Burkhardt, Richard. "Biology." In *Dictionary of the History of Science.* Edited by W. F. Bynum et al. London: Macmillan, 1981.

———. *The Spirit of System.* Cambridge: Harvard University Press, 1977.

Buttersack, Felix. "Karl Friedrich Kielmeyer." *Sudhoffs Archiv für Geschichte der Medizin und der Naturwissenschaften* 23 (1930): 236–46.

Caneva, K. L. "Teleology with Regrets." *Annals of Science* 47 (1990): 291–300.

Carpenter, William. *Principles of Comparative Physiology.* 4th ed. London: Churchill, 1854.

———. *Principles of General and Comparative Physiology.* London: Churchill, 1839.

Carus, Carl Gustav. *Lebenserinnerungen und Denkwürdigkeiten.* 4 vols. Leipzig: Brockhaus, 1865.

———. *Lehrbuch der Zootomie.* Leipzig: Gerhard Fleischer the Younger, 1818.

———. *Lehrbuch der vergleichenden Zootomie.* 2d ed. 2 vols. Leipzig: Fleischer, 1834.

[Chambers, Robert.] *Vestiges of the Natural History of Creation.* 3d ed. New York: Harper Brothers, n.d.

[_____.] *Vestiges of the Natural History of Creation.* 10th ed. London: Churchill, 1853.

Cicero. *De finibus bonorum et malorum.* 2nd ed. Loeb Classical Library, 1931.

Coleridge, Samuel Taylor. *Miscellanies, Aesthetic and Literary: To Which is Added the Theory of Life.* Edited by T. Ashe. London: Bell & Sons, 1892.

Corsi, Pietro. *The Age of Lamarck.* Berkeley: University of California Press, 1988.

Cunningham, Andrew, and Nicholas Jardine, eds. *Romanticism and the Sciences.* Cambridge: Cambridge University press, 1990.

Cuvier, Georges. "Eloge de M. De Lamarck." *Mémoires de l'Académie des sciences.* 2d ser., 13 (1835): i–xxxi.

_____. *Histoire des progrès des sciences naturelles depuis 1789 jusqu'à ce jour.* 4 vols. Paris: Baudouin Frères, 1829.

_____. *Le règne animal.* 2d ed. 5 vols. Vols. 4 and 5 by P. A. Latreille. Paris: Deterville, 1829–30.

_____. "Sur un nouveau rapprochement à établir entre les classes qui composent le règne animal." *Annales du Muséum d'Histoire Naturelle* 19 (1812): 73–84.

Darwin, Charles. *The Autobiography of Charles Darwin.* Edited by Nora Barlow. New York: Norton, 1969.

_____. *Charles Darwin's Notebooks, 1836–1844.* Edited by Paul Barrett, Peter Gautrey, Sandra Herbert, David Kohn, and Sydney Smith. Ithaca: Cornell University Press, 1987.

_____. *The Correspondence of Charles Darwin.* 6 vols. Cambridge: Cambridge University Press, 1985–.

_____. *The Descent of Man, and Selection in Relation to Sex.* 2 vols. London: Murray, 1871.

_____. *The Foundations of the Origin of Species: Two Essays Written in 1842 and 1844 by Charles Darwin.* Edited by Francis Darwin. Cambridge: Cambridge University Press, 1909.

_____. *Life and Letters of Charles Darwin.* Edited by Francis Darwin. 2 vols. New York: D. Appleton, 1891.

_____. *A Monograph of the Fossil Balanidae and Verrucidae of Great Britain.* London: Palaeontographical Society, 1854.

_____. *A Monograph of the Fossil Lepadidae or, Pedunculated Cirripedes of Great Britain.* London: Ray Society, 1851.

————. *A Monograph of the Sub-Class Cirripedia: The Balanidae (or Sessile Cirripedes), the Verrucidae, &c.* London: Ray Society, 1854.

————. *A Monograph of the Sub-Class Cirripedia, with Figures of all the Species: The Lepadidae or, Pedunculated Cirripedes.* London: Ray Society, 1851.

————. *More Letters of Charles Darwin.* 2 vols. Edited by Francis Darwin. London: Murray, 1903.

————. *On the Origin of Species.* London: Murray, 1859.

————. *The Origin of Species by Charles Darwin: A Variorum Text.* Edited by Morse Peckham. Philadelphia: University of Pennsylvania Press, 1959.

————. *The Red Notebook of Charles Darwin.* Edited by Sandra Herbert. Ithaca: Cornell University Press, 1980.

————. *The Variation of Animals and Plants under Domestication.* 2d ed. 2 vols. New York: D. Appleton, 1899.

Darwin, Erasmus. *Zoonomia or the Laws of Organic Life.* 2d ed. 2 vols. London: Johnson, 1796.

Desmond, Adrian. *The Politics of Evolution.* Chicago: University of Chicago Press, 1989.

Geoffroy Saint-Hilaire, Etienne. "Le degré d'influence du monde ambiant pour modifier les formes animals." *Mémoires de l'Académie royale des sciences de l'Institut de France,* 2d. ser. 12 (1833): 63–92.

————. *Principes de philosophie zoologique.* Paris: Didier, 1830.

————. "Sur une colonne vertébrale et ses cotes dans les insectes apiropodes." *Isis* 2 (1820): 527–52.

Glass, Bentley, ed. *Forerunners of Darwin.* Baltimore: Johns Hopkins University Press, 1968.

Goethe, Johann Wolfgang von. *Aus meinen Leben: Dichtung und Wahrheit.* 2d ed. Berlin und Weimar: Aufbau Verlag, 1984.

————. *Goethe, Die Schriften zur Naturwissenschaft.* 1st division, vols. 9 and 10, *Morphologische Hefte.* Edited by Dorothea Kuhn. Weimar: Böhlaus Nachfolger, 1954.

————. *Goethe, Die Schriften zur Naturwissenschaft.* 2d division, vols. 9a and 9b, *Zur Morphologie, von 1796 bis 1815.* Edited by Dorothea Kuhn. Weimar: Böhlaus Nachfolger, 1977–86.

————. "Metmorphose der Thiere." In *Goethes Gedichte in Zeitlicher Folge,* edited by Heinz Nicolai. 2d ed. Frankfurt am Main: Insel Verlag, 1982.

Gould, Stephen Jay. "Eternal Metaphors of Palaeontology." In *Pat-

terns of Evolution as Illustrated in the Fossil Record, edited by A. Hallan. New York: Elsevier, 1977.

—————. *Ever Since Darwin.* New York: Norton, 1977.

—————. *Ontogeny and Phylogeny.* Cambridge: Harvard University Press, 1977.

—————. *Wonderful Life: The Burgess Shale and the Nature of History.* New York: Norton, 1989.

[Grant, Robert.] "Lectures on Comparative Anatomy and Animal Physiology, Delivered during the Session 1833–4" (60 lectures). *Lancet* 1–2 (1833–34).

[—————.] "Observations on the Nature and Importance of Geology." *Edinburgh New Philosophical Journal* 1 (1826): 293–302.

Green, Joseph Henry. *Vital Dynamics: The Hunterian Oration before the Royal College of Surgeons in London, 14th February 1840.* London: Pickering, 1840.

Greene, John. *Science, Ideology, and World View.* Berkeley: University of California Press, 1981.

Haeckel, Ernst. *Generelle Morphologie der Organismen.* 2 vols. Berlin: Reimer, 1866.

Harvey, William. *Exercitationes de generatione animalium.* London: DuGaidianis, 1651.

Hodge, M. J. S. "Darwin and the Laws of the Animate Part of the Terrestrial System (1835–1837)." *Studies in the History of Biology* 6 (1983): 1–106.

—————. "Darwin as a Lifelong Generation Theorist." In *The Darwinian Heritage,* edited by David Kohn. Princeton: Princeton University Press, 1985.

Huxley, Thomas Henry. *Evidence as to Man's Place in Nature.* London: Williams and Norgate, 1863. Photoreproduction. Ann Arbor: University of Michigan Press, 1959.

—————. "Evolution." *Encyclopaedia Britannica.* 9th ed. New York: Charles Scribner's Sons, 1878.

—————. *Life and Letters of Thomas Henry Huxley.* Edited by Leonard Huxley. 2 vols. New York: Appleton, 1900.

—————. "The Natural History of Creation." *Academy* 1 (1869): 13–14, 40–43.

—————. "On the Morphology of the Cephalous Mollusca." In *The Scientific Memoirs of Thomas Henry Huxley,* edited by Michael Foster and E. Ray Lankester. 4 vols. London: Macmillan, 1898.

[—————.] "Science." *Westminster Review* 63 (n.s., 7) (1855): 239–53.

————. "Vestiges of the Natural History of Creation, Tenth Edition." In *The Scientific Memoirs of Thomas Henry Huxley,* edited by Michael Foster and E. Ray Lankester. Supplement. London: Macmillan, 1898.

Iggers, Georg. The German Conception of History. 2d ed. Middletown, Conn.: Wesleyan University Press, 1983.

Kant, Immanuel. *Kritik der Urteilskraft.* In *Immanuel Kant Werke,* edited by Wilhelm Weischedel. 6 vols. Wiesbaden: Insel-Verlag, 1957.

Kielmeyer, Karl Friedrich. "Ueber die Verhältnisse der organischen Kräfte unter einander in der Reihe der verschiedenen Organisation" (1793). Reprinted in *Sudhoffs Archiv für Geschichte der Medizin und der Naturwissenschaften* 23 (1930): 247–67.

Kohn, David. "Darwin's Principle of Divergence as Internal Dialogue." In *The Darwinian Heritage,* edited by David Kohn. Princeton: Princeton University Press, 1985.

————. "Theories to Work by: Rejected Theories, Reproduction, and Darwin's Path to Natural Selection." *Studies in the History of Biology* 4 (1980): 67–170.

Kovalevsky, A. "Entwicklungsgeschichte der einfachen Ascidien." *Mémoires de l'Académie impériale des sciences de St.-Pétersbourg,* 7th ser., vol. 10, no. 15 (1866).

————. "Weitere Studien über die Entwicklung der einfachen Ascidien." *Archiv für Mikroskopische Anatomie* 7 (1871): 101–30.

Lamarck, Jean-Baptiste de. *Philosophie zoologique.* 2 vols. Paris: Dentu, 1809.

————. *Système des animaux sans vertèbres.* Paris: Lamarck et Deterville, 1801.

Lenoir, Timothy. *The Strategy of Life.* Chicago: University of Chicago Press, 1989.

Lyell, Charles. *Principles of Geology.* 3 vols. London: Murray, 1830–33.

Lyons, Sherrie. "The Evolution of Thomas Henry Huxley's Evolutionary Views." Ph.D. diss., University of Chicago, 1990.

Mayr, Ernst. *The Growth of Biological Thought.* Cambridge: Harvard University Press, 1982.

Meckel, Johann Friedrich. *Abhandlungen aus der menschlichen und vergleichenden Anatomie und Physiologie.* Halle: Hemmerde und Schwetschke, 1806.

————. *Beyträge zur vergleichenden Anatomie.* 2 vols. Leipzig: Reclam, 1808–12.

_____. *System der vergleichenden Anatomie.* Vol. 1. Halle: Renger, 1821.

Mill, John Stuart. *John Stuart Mill, On Bentham and Coleridge.* Introduction by F. R. Leavis. New York: Harper Torch, 1962.

Millhauser, Milton. *Just Before Darwin: Robert Chambers and Vestiges.* Middletown, Conn.: Wesleyan University Press, 1959.

Milne-Edwards, Henri. "Considérations sur quelques principes relatifs à la classification naturelle des animaux." *Annales des sciences naturelles,* 3d ser., 1 (1844): 65–99.

Müller, Johannes. *Elements of Physiology.* 2 vols. Translated by William Baly. London: Taylor & Walton, 1837–42.

Nyhart, Lynn. *Before Biology: Animal Morphology and the German Universities, 1850–1900.* Forthcoming.

Oken, Lorenz. *Abriss des Systems der Biologie.* Göttingen: Vandenhoek und Ruprecht, 1805.

_____. *Allgemeine Naturgeschichte für alle Stände.* 7 vols. Stuttgart: Hoffman, 1833–41.

_____. *Die Zeugung.* Bamberg und Wirzburg: Goebhardt, 1805.

_____. *Lehrbuch der Naturphilosophie.* 2d ed. Jena: Frommann, 1831.

_____. *Ueber die Bedeutung der Schädelknochen.* Bamberg: Göbhardt, 1807.

_____. "Ueber die Bedeutung der Schädelknochen." *Isis oder Encyclopädische Zeitung* 1 (1817): 1204–9.

Oppenheimer, Jane. "An Embryological Enigma in the *Origin of Species.*" In *Forerunners of Darwin,* edited by Bentley Glass. Baltimore: Johns Hopkins University Press, 1968.

Ospovat, Dov. *The Development of Darwin's Theory.* Cambridge: Harvard University Press, 1981.

[Owen, Richard.] "Darwin on the Origin of Species." *Edinburgh Review* 11 (1860): 487–532.

[_____.] *Descriptive and Illustrated Catalogue of the Physiological Series of Comparative Anatomy contained in the Museum of the Royal College of Surgeons in London.* Vol. 5, *Products of Generation.* London: Taylor, 1840.

_____. *Lectures on the Comparative Anatomy and Physiology of the Invertebrate Animals, Delivered at the Royal College of Surgeons in 1843.* Notes taken by William White Cooper and revised by Richard Owen. London: Longman, Brown, Green, and Longmans, 1843.

_____. *Lectures on the Comparative Anatomy and Physiology of the*

Vertebrate Animals, Delivered at the Royal College of Surgeons of England in 1844 and 1846. London: Longman, Brown, Green, and Longmans, 1846.

[————.] "Lyell—on Life and its Successive Development." *Quarterly Review* 89 (1851): 412–51.

————. "On British Fossil Reptiles." *Edinburgh New Philosophical Journal* 33 (1842): 65–88.

————. *On the Nature of Limbs.* London: Van Voorst, 1849.

————. "Remarks on the Entozoa." *Transactions of the Zoological Society of London* 1 (1835): 387–94.

————. "Report on the Archetype and Homologies of the Vertebrate Skeleton." *Report of the Sixteenth Meeting of the British Association for the Advancement of Science: Held at Southampton in September 1846,* pp. 169–340. London: Murray, 1847.

————. *Richard Owen's Hunterian Lectures, May–June 1837.* Edited by Phillip Sloan. London: British Museum (Natural History); Chicago: University of Chicago Press, forthcoming.

"Review of *Historia Generalis Insectorum,*" *Philosophical Transactions of the Royal Society* 5 (1670): 2078–80.

Richards, Evelleen. "A Question of Property Rights: Richard Owen's Evolutionism Reassessed." *British Journal for the History of Science* 20 (1987): 129–71.

Richards, Robert J. "Christian Wolff's Prolegomena to Empirical and Rational Psychology: Translation and Commentary." *Proceedings of the American Philosophical Society* 124 (1980): 227–39.

————. *Darwin and the Emergence of Evolutionary Theories of Mind and Behavior.* Chicago: University of Chicago Press, 1987.

————. "Influence of Sensationalist Tradition on Early Theories of the Evolution of Behavior." *Journal of the History of Ideas* 40 (1979): 85–105.

————. "The Moral Foundations of the Idea of Evolutionary Progress: Darwin, Spencer, and the Neo-Darwinians." In *Evolutionary Progress,* edited by Matthew Nitecki. Chicago: University of Chicago Press, 1988.

Richmond, Marsha. "Darwin's Study of Cirripedia." In Charles Darwin, *The Correspondence of Charles Darwin,* 4: 388–409. Cambridge: Cambridge University Press, 1985–.

Roe, Shirley. *Matter, Life, and Generation: 18th-Century Embryology and the Haller-Wolff Debate.* Cambridge: Cambridge University Press, 1981.

Roger, Jacques. *Les sciences de la vie dans la pensée française du XVIIIe siècle.* 2d ed. Paris: Armand Colin, 1971.

Ruse, Michael. *The Darwinian Revolution.* Chicago: University of Chicago Press, 1979.

———. "Molecules to Men: Evolutionary Biology and Thoughts of Progress." In *Evolutionary Progress,* edited by Matthew Nitecki. Chicago: University of Chicago Press, 1988.

Russell, E. S. *Form and Function: A Contribution to the History of Animal Morphology.* London: Murray, 1916. Reprint. Chicago: University of Chicago Press, 1982.

Schelling, Friedrich Wilhelm von. *Sämmtliche Werke.* Edited by K. F. A. Schelling. 14 vols. Stuttgart and Augsburg: Cotta'scher Verlag, 1856–61.

Secord, James. "Behind the Veil: Robert Chambers and *Vestiges.*" In *History, Humanity, and Evolution,* edited by James Moore. Cambridge: Cambridge University Press, 1989.

———. "Edinburgh Lamarckians: Robert Jameson and Robert E. Grant." *Journal of the History of Biology* 24 (1991): 1–18.

Serres, Etienne Reynaud Augustin. *Anatomie comparée du cerveau.* 2 vols. Paris: Gabon, 1827.

———. "Théorie des formations organiques: Formes transitoires et permanentes des organes." *Annales des sciences naturelles* 12 (1827): 82–104.

———. "Zoologie: Anatomie des mollusques." *L'Institut, section des Sciences mathématiques, physiques et naturelles,* no. 191 (4 January 1837): 370–71.

Sloan, Phillip. "Buffon, German Biology, and the Historical Interpretation of Biological Species." *British Journal for the History of Science* 12 (1979): 109–53.

Spencer, Herbert. *Social Statics, or the Conditions Essential to Human Happiness.* London: Chapman, 1851.

Sulloway, Frank. "Darwin's Conversion: The *Beagle* Voyage and Its Aftermath." *Journal of the History of Biology* 15 (1982): 325–96.

Swammerdam, Jan. *Historia insectorum generalis.* Translated by H. Henninius. Holland: Luchtmans, 1685.

Tiedemann, D. Friedrich. *Anatomie und Bildungsgeschichte des Gehirns im Foetus des Menschen nebst einer vergleichenden Darstellung des Hirnbaues in den Thieren.* Nuremberg: Steineschen Buchhandlung, 1816.

————. *Zoologie, zu seinen Vorlesungen Entworfen.* 3 vols. Land-shut: Weber, 1808–14.

Treviranus, Gottfried Reinhold. *Biologie, oder Philosophie der le-benden Natur.* 6 vols. Göttingen: Johann Friedrich Röwer, 1802–22.

Vogt, Carl. *Embryologie des salmones.* Vol. 2 of Louis Agassiz, *Histoire naturelle des poissons d'eau douce.* Neuchâtel: Petitpierre, 1842.

Whewell, William. *History of the Inductive Sciences.* 3 vols. London: Parker, 1837.

Whitman, Charles Otis. "Bonnet's Theory of Evolution." In *Biological Lectures Delivered at the Marine Biological Laboratory of Wood's Hole,* edited by C. O. Whitman. Boston: Ginn & Company, 1895.

————. "The Palingenesia and the Germ Doctrine of Bonnet." In *Biological Lectures Delivered at the Marine Biological Laboratory of Wood's Hole,* edited by C. O. Whitman. Boston: Ginn & Company, 1895.

Wilson, Leonard. *Charles Lyell: The Years to 1841.* New Haven: Yale University Press, 1972.

Wolff, Caspar Friedrich. *Theoria Generationis.* Halle: Hendelianis, 1759.

————. *Theorie von der Generation.* Berlin: Birnstiel, 1764.

Wolff, Christian. "Prolegomena to Empirical and Rational Psychology." Translation and commentary by Robert J. Richards. *Proceedings of the American Philosophical Society* 124 (1980): 227–39.

Young, Robert. "Darwinism is Social." In *The Darwinian Heritage,* edited by David Kohn. Princeton: Princeton University Press, 1985.

INDEX